THE QUANTUM SOCIETY

THE QUANTUM SOCIETY

THE QUANTUM SOCIETY

THE QUANTUM SOCIETY

THE QUANTUM SOCIETY

THE QUANTUM SOCIETY

THE QUANTUM SOCIETY

THE QUANTUM SOCIETY

THE QUANTUM SOCIETY

THE QUANTUM SOCIETY

THE QUANTUM SOCIETY

THE QUANTUM SOCIETY

THE QUANTUM SOCIETY

Also by Danah Zohar

Through the Time Barrier: A Study of Precognition and Modern Physics
The Quantum Self

# THE
# QUANTUM
# SOCIETY

*Mind, Physics, and a*
*New Social Vision*

**DANAH ZOHAR**
**& IAN MARSHALL**

*This book
is nearly
20 yrs.
old.*

*Quill · William Morrow · New York*

Library of Congress Cataloging-in-Publication Data

Zohar, Danah, 1945-
  The quantum society: mind, physics, and a new vision  /  by
Danah Zohar and Ian Marshall.
  p. cm.
   ISBN 0-688-14230-3
   1. Social psychology. 2. Quantum theory. 3. Physics—
Philosophy.
  1. Marshall, I. N. II. Title.
HM291.Z64 1994             93-34348
302–dc20                CIP

Printed in the United States ot America


  5  6  7  8  9  10

BOOK DESIGN BY LINEY LI

*To the memory of David Bohm,*

*a good friend and a pioneer in this field*

See page 166 for an excellent commentary of collapse and black holes.

See pages 232-233 for science → sociology comparative for theme of the whole book.

See pp. 288-90 on the failures in British government!

See pp. 305-320 on marriage, sex and family. (Excellent).

Ch. 13 should be would required reading!

## • *Preface* •

This book developed largely through the response of others. After publication of *The Quantum Self,* hundreds of people wrote to me, others spoke to me after meetings or conferences I had been asked to address. These people were excited by the vision of a self that is deeply at one with fundamental reality and essentially defined through its relationship to others and to the natural world, but they wanted to know how they could take it further.

Some asked how the model for intimate relationship that we find in quantum reality could be extended to larger groups in society. Others thought they saw a model for getting beyond the politics of conflict and confrontation in the way that quantum systems replace collision with a tendency to overlap and combine when they meet. A group of management students wondered whether quantum indeterminacy applied to human behavior didn't relate to their search for a more flexible, less hierarchic management structure. The same kind of question arose about discovering a more flexible response to the increasing ambiguity surrounding role and relationship in society at large. One Oxford undergraduate even asked me how we could build "a quantum society."

These questions came at the right time for me. Writing *The Quantum Self* had been a very personal experience. It had changed my whole perception of myself and my place in the general scheme of things. Seeing my own life as part of an ongoing process of evolution in the universe and as something that continues through its relationship to others, I overcame a lifelong fear of death. Sensing the self's deeper roots in a creative dialogue with the whole of reality left me with a strong sense that my life wasn't solely my own concern.

I, too, wanted to take these insights further. My growing point had shifted to a larger stage and reawakened something in me that I thought I had lost forever. That something was my whole sense of myself as a member of wider society.

As a child in the United States I had understood myself almost wholly in terms of being "an American." This sense of being an American can be, as many know, almost mystical. It is associated with a vision of the good life, a vision of how one relates to one's fellow Americans and to the world at large, a deep sense of the meaning and importance of one's actions so that these adequately reflect American values and bring credit to one's nation and family. My early teenage interest in science was awakened as much by an impulse to serve America as by a fascination with atomic physics. Even my religious feelings were inseparable from a perhaps naive, but not untypical, belief that in serving America I would be serving God. This same quasi-mystical dimension is associated with the nationality of many nations.

When I was in my late teens and early twenties I lost all this. I was nineteen when President John Kennedy was assassinated. As for so many others of my generation, that death did violence to much that I associated with being young and being American. The president's assassination was followed in rapid succession by the murders of Martin Luther King and Bobby Kennedy, and by America's deep involvement in the Vietnam War. Some dark and irrational force or forces seemed to me bent on destroying all that was best in my society and what it stood for, and the whole edifice of my childhood beliefs came tumbling down. It was impossible to reconcile my youthful ideals with events unfolding both at home and abroad.

My own reaction to America's political traumas of the sixties was to leave. I became an expatriate and lived as a stranger in foreign lands for the next twenty-five years. I first changed my religion to Judaism and moved to Israel, where to some extent I took perverse comfort in being a stranger even to myself. I took no trouble to learn the language of my host nation and satisfied myself with the social distance imposed by an inability to understand the conversations of one's fellows. Even when I moved to England four years later and made it my home, I lived as an outsider who made no attempt to involve myself in British society. I married an Englishman who was himself in many ways a social outsider and we collected around ourselves others who were expatriates or who lived on the fringes of society. For twenty years I did not visit America, even for the funerals of my mother and grandmother.

With the birth of my two children and the completion of *The Quantum Self,* I was no longer satisfied with my life in exile. The book had brought me closer to my immediate family and I now felt a need to see that family as a part of wider society. I felt a need to see myself as part of wider society, to find some social role. I even found a renewed interest in politics and wished for some authentic way to feel a part of the political process. At the same time, my now deeper spiritual sense had taken a more social direction and I longed to express it in the company of friends and community.

These newly reawakened needs were exciting, but they also evoked a sense of frustration, a frustration that I believe many of us share. It is a frustration that raises deep questions about the whole nature of contemporary society and social involvement. What kind of social role, for instance, would satisfy a wish for creative involvement in today's society? Indeed, in what kind of society would we want to become involved? How is it possible to be an authentic participant in the existing political process? In what sort of familiar community structure could we possibly share the kind of spontaneous, heterodox, and often earth-centered spirituality that so many are feeling? Simply raising such questions exposes the obvious lack of good answers within the old social, political, and religious structures.

*The Quantum Self* suggested a new way forward for the creative individual in his or her personal life. It drew on the insights and principles of quantum reality to lay a new foundation for human consciousness and the nature of the self. Creativity itself was linked to creative processes in the universe as a whole, and the self that emerged was a self deeply rooted in the fundamental forces of relationship that bind the universe together and cause it to evolve.

In this book I want to extend that quantum model of the person to a more all-embracing quantum model of society. I want to discover the roots of a more responsive, less hierarchic society— a more *natural* society—in the basic structures and dynamics of fundamental reality from which the dynamics of human groups emerge. I believe this is necessary not just to understanding the true nature of society, and thus being in a position to transform existing patterns and institutions. It is also necessary to any full appreciation of what it means to be a self within society.

Until a few centuries ago, the "self" was not even mentioned. People were units of society. They saw themselves mirrored in the social roles and structures of society and its duties. With the rise of individualism in the seventeenth century, the balance tipped radically in the opposite direction. The self, with its inner needs and desires and its claim to rights, emerged as all-important. Society was just a neutral stage on which the self played out its private drama. This emphasis contributed to much of the isolation and fragmentation that afflict modern life and led inevitably to experiments in renewed collectivism. But neither extreme answers to the needs of today.

The main challenge of our times is to link the inner world of the self with the outer world of society, and to see both within the larger context of the natural world. To do this effectively, I believe we must come to appreciate that self, society, and nature all derive from a common source, that each is a necessary partner in some larger creative dialogue.

I often meet people who ask me what I write about. My answer that I write about physics and human nature or physics and society inevitably evokes expressions of surprise. To most people, the world of physics seems remote. Its abstract mathematical formulae and complex experimental results seem to bear no relation to the concerns of everyday life, to the passions that we feel, the kinds of decisions that we must make from moment to moment, to the nature of our social institutions. Yet we human beings are physical creatures. The dynamics of both our bodies and our minds emerge from the same laws and forces that move the sun and the moon or that bind atoms together.

There is just one reality, and we are all part of it. Physics tells us about the processes of creativity and transformation in the natural world. The physics of consciousness tells us about those same processes within ourselves and our society. If we understand the actual physical basis of transformation, perhaps we can align ourselves with it. Perhaps we can help it to unfold more quickly, with fewer mistakes.

Still, to speak of physics in the same context that we speak of something like consciousness, self-development, or social transformation will confound most people who are accustomed to thinking of science, and particularly of physics, as a force that has tended to dehumanize and demystify so much of our experience. Surely it is the "alien and inhuman world that science presents for our imagination," as the British philosopher Bertrand Russell described it, that has led to an "unyielding despair" about all human purposes and projects, personal as well as social and spiritual.[1] But the science of which Russell wrote, and the whole worldview associated with that science, is not the radically different science of the twentieth century. The physics of Isaac Newton is now known to be a limited approximation valid only within a narrow range of our experience. It has been superseded primarily by quantum physics and, more recently still, by the exciting new physics of chaos and complexity theory.

It was a central argument of *The Quantum Self*, and remains so for this book, that there is a whole rich repository of language, metaphor, and allusion in these new scientific ideas, as well as practical applications for understanding human nature and consciousness. Quantum physics in particular almost cries out for use as a more general model for a whole new kind of thinking about ourselves and our experience. There is an uncanny and intriguing similarity between the way that quantum systems relate and behave and so much that we are now beginning to understand or hope for about human social relations. Indeed, many of the ideas presented in this book reflect similar attempts by thinkers in many other fields to express a new vision for human society.

While I was writing this book, for instance, I either met or read the work of others from widely different disciplines and areas of interest whose concerns and conclusions about some of the larger social, political, or spiritual challenges facing us uncannily mirrored some of my own. I think particularly of ecofeminist Charlene Spretnak, sociologists Peter Berger and Richard Sennett, international lawyer Richard Falk, political writers David Osborne and Ted Gaebler, theater director Peter Brook, architect Richard Rogers, the Christian theologians David Ray Griffin and Father Thomas Berry, and the chief rabbi of Great Britain, Dr. Jonathan Sacks. It was exciting to integrate their insights with my own similar ones arising from the completely different standpoint of physics. Doing so made it all the more clear to me that there is genuinely a new paradigm emerging that is being articulated in different ways in the characteristic languages of diverse approaches. I hope that *The Quantum Society* will help to make clear the central features and implications of this new paradigm and what each of us can do as individuals to realize its practical application in our personal and social lives.

I would like to thank Charles Jencks for bringing me together with some of the above in a seminar that was heady indeed and to which he contributed so much. I would also like to thank the Athenian Society of Greece and Brahma Kumaris University for organizing the excellent Second International Symposium on Science and Consciousness in which I participated and through which I understood a great deal about the nature of dialogue. I am grateful to Colonel Mike Dewar of the Institute for Strategic Studies for his briefing on the thinking that underlies defense analysis, to Dr. Tali Loewenthal for his quite different insights into Jewish mystical thought, and to Rabbi Shmuley Boteach and his wife Debbie for sharing the blessings of their spiritual community. Once again, my editor at William Morrow, Maria Guarnaschelli, has contributed her invaluable support, suggestions for improvement, and "finishing touches." The book is much better for her contribution. With my good friends Layla Shamash and Khalil Norland I have enjoyed the most appositely stimulating conversations.

My special gratitude, as always, goes to my husband, Dr. Ian

Marshall. I wrote this book, but he contributed at least half of the metaphors and structure and did much of the research, and most of the ideas themselves usually evolved in conversations between us. Chapters 3 and 7 are almost wholly his work, as is the thinking in Chapters 11 and 12 on small and large groups. He is in every meaningful sense the book's coauthor.

I want to comment briefly on the level at which this book is written. Nearly all of it should be accessible to the general reader who has no deep knowledge of science or of any of the scientific issues raised. Indeed it has been written especially to excite the imaginations of such readers. The possible exception to this is Chapter 3, on the new physics of the mind. Some of the material in this chapter is difficult and may, admittedly, have more meaning to those already familiar with some of the issues at stake in the current debate on the nature of mind. For those able to follow its argument, I think the chapter adds a vital level of strength to the book's whole thesis. But I hope that the general reader will not get bogged down in this chapter (or fear needlessly that the whole rest of the book will be as difficult) and indeed that he or she may gain much positive understanding about the possible link between quantum phenomena and brain processes. The main point of the chapter is simply to argue that there are good reasons for supposing that mind actually has a quantum dimension and that therefore our social potential might derive, literally, in part from this.

Finally, I want to say something about the book's scope. It is clear that many volumes could be written on the problems facing contemporary society. Each of the chapters here could, and perhaps should, be a book in itself. I have tried as much as possible to restrict myself in two ways. I wanted, on the one hand, to concentrate on those aspects of our social and political challenge where the individual can make a difference, where some sort of personal transformation might in itself contribute to social transformation. Second, because so much of the thinking here is new and offers the possibility of an all-embracing new social model, I wanted to give an overview of the material. That, perhaps unfortunately, has at times meant that I merely touch on issues that cry out for more lengthy discussion. Readers who want to pursue any of these matters further can do so to some extent by reading the work of others I have just mentioned or by following up some of the specific references mentioned in the notes. I hope that any such further research will then make deeper sense still in the light of the overarching vision I have tried to offer here.

# • Contents •

*At the still point of the turning world.*

*Neither flesh nor fleshless;*

*Neither from nor towards;*

*at the still point, there the dance is,*

*. . . and there is only the dance.*

T. S. ELIOT
*Four Quartets*

# 1. What Is a "Quantum Society"?

*The weapons of the positive revolution are not bullets and bombs but simple human perceptions. Bullets and bombs may offer physical power but eventually will only work if they change perceptions and values. Why not go the direct route and work with perceptions and values?*

Edward de Bono*

We can think of society as a milling crowd, millions of individuals each going his or her own way and managing, somehow, to coordinate sometimes. This is the Western way.

We can think of society as a disciplined army, each member a soldier marching in tight, well-ordered step. Individual differences are suppressed for the sake of uniform performance. This is the now-discredited collectivist way.

Or, we might think of society as a free-form dance company, each member a soloist in his or her own right but moving creatively in harmony with the others. This is the new way I am going to describe in this book.

This is a book about changing our social perceptions and values. It is about changing the way we look at society and at our relations to each other within it. And it is a book about restructuring the customs and institutions of society through the power of this new vision. It is, if you like, a book about learning to dance together.

*Handbook for the Positive Revolution.

It is also, though, a book about fundamental physical reality, about the latest insights of quantum physics and quantum thermodynamics and how these insights can relate to our everyday concerns about self and society. In particular, I shall be looking at the origins of human consciousness in the wider world of the new physics and suggesting how we might use an understanding of those origins to define a new social vision that is at one with ultimate reality.

The idea of a "quantum society" stems from a conviction that a whole new paradigm is emerging from our description of quantum reality and that this paradigm can be extended to change radically our perception of ourselves and the social world we want to live in. I believe that a wider appreciation of the revolutionary nature of quantum reality, and of the possible links between quantum processes and our own brain processes, can give us the conceptual foundations we need to bring about a "positive revolution" in society.

The notion of a quantum society is also meant to suggest the image of a society firmly rooted in nature. Rooted not just in the nature of trees and rivers and the ecosphere, but in the nature of physical reality itself—a society drawing its laws and principles, its self-images and its metaphors from the same laws and principles underlying all else that is in the universe.

One of the leading physicists of this century, David Bohm, has used the image of a dance to illustrate the dynamic unity of quantum laws and principles. He compares the movements of electrons in the laboratory to those of ballet artists responding to a musical score. The score itself, he says, constitutes "a common 'pool' of information" through which the dancers can move together in an organized and orderly way.[1]

The human brain is the natural link between our perceptions and values and the "cosmic dance" of physical reality. The brain is the physical basis of our conscious life. A better understanding of how the brain gives rise to mind can give us a better understanding of the potential latent within thought. Early on in this book, then, I am going to suggest a new quantum physical model of how the brain works and explore how the use of such a model can itself become a powerful "weapon" for personal and social transformation.

The word *society* is used in many ways to describe quite different social arrangements—nation-states such as "British society" or "American society," religious or ethnic subcultures such as "Christian society," or even the cultural patterns of class groups such as "middle-class society" or "high society." I am writing here about society in its most general and inclusive sense, as the domain in which we dwell together with others. This domain, which includes what sociologist Emile Durkheim calls our "collective patterns of thinking, feeling and action," extends from the most private world of intimate relationships to the increasingly global world of economic, political, and power relationships. I assume that to be involved in relationship, or even impinged upon by it, is to dwell in society.

It is nothing new to suggest a direct link between our understanding of the physical world and our thinking about patterns of social and political relationship. The ancient Greeks believed that the movements of the heavens were the source of all transformation in the world, social as well as physical. Heraclitus' principles of strife and tension in the physical realm were applied to relations between people in society. The vortex theories of Democritus and Epicurus were an early Greek attempt at thermodynamics. They saw the earth at the center of a vast, eddying whirlpool of ether that kept the stars and planets in motion. These physical images or visions also became the central explanatory principle for all social change.[2]

Today, our perception of social and political reality, our whole perception of "modernity," is a mechanistic perception. It was formed in direct response to the philosophical and scientific revolution of the seventeenth century that gave birth to modern science and is reinforced daily by our constant exposure to the technology that surrounds us. The greatest figure in this new mechanical science, Isaac Newton, believed that the foundations of his work could be applied to problems in moral philosophy.[3] The French Polytechnicians of the eighteenth century tried to extend his ideas to questions of history and spirit. As Giorgio de Santillana

This is my favourite Self Portrait

**FIG. 1.1  THE MECHANISTS' VISION OF THE HUMAN MACHINE**

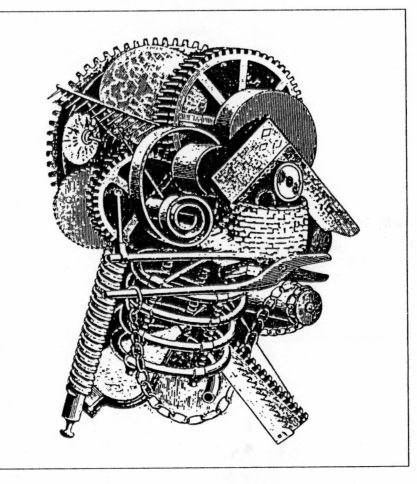

expressed it, "these men tried to build a religion as they had learned to build a bridge."[4] Others shared this wider mechanical vision. The sheer power and simplicity of Newton's three mechanical laws of motion, and the apparent force of the new empirical method, drew nearly every influential social, political, and economic thinker of the seventeenth, eighteenth, and nineteenth centuries to use them as a model.

Thomas Hobbes,* John Stuart Mill, and even John Locke

---

*Hobbes published *Leviathan* before Newton had written the *Principia,* but many of the mechanical ideas formalized by Newton were circulating before through the work of Kepler, Galileo, and others.

turned to the new mechanical physics for inspiration and example in their writings on state and society.[5] Locke once described himself as a "mere underlabourer" to the "incomparable Mr. Newton." Adam Smith, too, was very impressed by the new mechanistic science and modeled both his free-market economics and the division of labor on Newton's methods.[6] Marx's deterministic laws of history, Darwin's mechanistic and reductionist theory of evolution, and Freud's "scientific" model of the self as a complex hydraulic system issue from the same source. Among the early pioneers of modern sociology, Vilfredo Pareto drew heavily on mechanical and thermal metaphors when describing the dynamics of society.[7] The man who invented the word *sociology* itself, Auguste Comte, first called his new science a "social physics."

Where Newton formulated the fundamental laws of physical reality, the philosophers and sociologists following in his wake hoped to discover the basic axioms and principles of social life. His universal clockwork machine became their model for comparing the state to a precise, law-abiding mechanism and portraying human beings as living machines (Fig. 1.1). Both metaphors survive today in expressions like "the wheels of government" and "the machinery of state" and in the philosophical implications of artificial intelligence—we are "mind machines,"[8] we "switch on" and "switch off," we "blow our fuses" and are "programmed" for success or failure.

The basic building blocks of Newton's physical world were so many isolated and impenetrable atoms that bounce around in space and collide with one another like tiny billiard balls. The *only* actors in Newton's space-time drama were such particles and the attractive or repulsive forces acting between them. Political thinkers of the time compared these colliding atoms and their interacting forces to the behavior and interactions of individuals in society as they confront each other in pursuit of their self-interest.[9] In the *Leviathan*, Thomas Hobbes likened it to a "war of every man against every man."[10] Still today, economists and sociologists who follow "Rational Choice" theory argue that individuals will always choose to act in pursuit of their own self-interest.[11] Finding some way to balance all the conflicting interests that result in society has been the basis for adversarial democracy and the familiar confrontational style of modern political parties.[12]

Mechanistic physics became the touchstone for a whole worldview, the central paradigm of the modern world. In many ways it served us well. It made possible our whole industrial and technological culture and was intellectually compatible with the flowering of both Western individualism and liberal democracy. But the extension of the mechanistic paradigm to our general perception of reality, both physical and social, has had consequences we are now beginning to question.

Mechanism stresses an unbridgeable gulf between human beings and the physical world. Human consciousness has no role or place in Newton's vast world machine. As the French biologist Jacques Monod describes it, we live "like gypsies . . . on the boundary of an alien world."[13] This sense of an alien physical realm was extended, in association with Christian influence, to the wider world of nature. Nature is perceived as wholly "other" than ourselves, a force to be conquered and used. We can see in such distorted perception the origins of our current ecological crisis.

Mechanism stresses the absolute, the unchanging, and the certain. Ambiguity is its enemy. Newton's absolute space-time coordinates are the framework for a fixed, predictable, and rigidly law-abiding universe. Mechanistic society stresses the absolute center, with power radiating outward. It stresses fixed role-playing and rigid bureaucratic organization.

Mechanism stresses hierarchy. It structures existence according to ever-descending units of analysis. Molecules are more basic than neurons, atoms more basic than molecules. We structure power and organization in the same ladder of ascending and descending authority.

Mechanism stresses isolated, separate, and interchangeable parts. Everything in Newton's universe is ultimately reducible to so many individual atoms and the forces acting between them. Atomism encourages a model of relationship based on conflict and confrontation, on part against part. In our times Hobbes's mechanistic "war of every man against every man" takes the literal form of ultimate conflict. "Most obviously," says Princeton's Richard Falk, "nuclear weapons as instruments for struggle by part against part doom the whole and overwhelm any possibility of modernist sanity."[14]

Atomism underlies the modern cult of the expert, the detached individual who is very knowledgeable about isolated bits of information or experience but ignorant of the whole of which these bits are a part. The parts are alienated from each other and from the whole; the whole is subject to fragmentation. The expert is alienated from the situation or community in which he practices his expertise.

The Industrial Revolution, and the mass production that followed in its wake, extended this alienation to our very understanding of human beings and the nature of our labor. In the vast industrial machine (or the large corporate organization), the individual laborer becomes a "factor of production," an objectified unit in the standardized production process. His personal and social relationships, and anything we might define as his spiritual qualities, are isolated from the wholly separate and rigidly bureaucratized world of work. Mechanism's employees are, as Marx pointed out, alienated both from themselves as wider beings and from the products of their own labor.

Mechanism stresses the single point of view. In an absolute space-time framework there is only one way of looking at things. In Newtonian physics, there is only one reality at a time. The either/or of absolute choice becomes the favored way of dealing with reality. A statement is either true or false, a course of action is either good or bad. There can be only one truth, only one best course of action. Neither nuance nor paradox, neither multiplicity and difference nor plurality can be accommodated.

Mechanism models itself on the early low-tech machine. The universe is made up of fixed working parts, their laws of interaction strictly determined. Such machines are heavy, unresponsive things. They can't cope with change. They leave no scope for flexibility. The "machinery of government" and the "cogs and wheels of organization" suffer the same drawbacks. Excessive bureaucracy, which treats people as faceless units to be dealt with by impersonal rules, increasingly is seen to be both alienating and inefficient.[15]

In science itself, mechanism has long since had its day. Newton's strict determinist laws still apply within a narrow spectrum of physical reality, but they are no longer at the cutting edge of physical thinking. As a more general model of reality, too, mechanism now strikes us as painfully limited. In the words of Dominican writer Thomas Berry, we now appreciate that under its influence, "the human mind lived in the narrowest bonds that it has ever experienced."[16]

The failure to give any account of where life and consciousness belong in the universe leaves human beings with no sense of our place in the scheme of things. Atomism denies the reality and importance of relationship, establishing a precedent for conflict and confrontation and the pursuit of limiting self-interest. As a model, mechanism cannot account for why people ever act on behalf of others, or why there is any sort of social cohesion. And its sharp separation between the mental and the physical encourages a division between ourselves and the natural world, setting us in opposition to the world of nature and to the natural within ourselves.

But as futurist writer Alvin Toffler expresses it, "The Age of the Machine is screeching to a halt."[17] Its image no longer nourishes us, it no longer works as a symbol that binds our culture together and gives it meaning.

New patterns of social and political relationship are at the forefront of our attention today. Familiar patterns of family, community, business, and even nation are breaking down. Old political and economic systems are crumbling or straining to the breaking point. They no longer answer to our deepest needs and questions, they no longer inspire us with a vision or motivate us to action. Our relationship to nature has reached a point of global urgency. The whole mechanistic paradigm of society can no longer cope with contemporary reality.

At the same time, something new is "in the air," a new emphasis on unity and integration, talk of a new politics, a new society, a new spirituality. In both our private lives and our public roles we are seeking a whole new *framework* for understanding and fulfilling our potential as social beings. We are seeking ways to articulate and to institutionalize a new kind of social reality.

Because it is a social reality, it must be a *shared* reality. The exhaustion of traditional social and political patterns, and the helter-skelter juxtaposition of knowledge and experience from a whole panoply of once distinct and locally rooted cultures, has left us with the characteristic fragmentation of modern life. We lack a consensus about meanings, values, customs, and "symbols." We have lost what sociologists call the "taken-for-grantedness" of social reality.[18] We must learn to reexperience this reality as an integrated whole.

There are several features that we know this new social reality must have:

• *It Must Be "Holistic."* Globally, we live in a world of growing social, political, and economic interdependence. Even small shifts within society or within some one part of society are felt throughout the world. Stock market jitters in Tokyo are reflected in London and New York within hours. A revolution in Beijing raises hope and expectation in Europe and the Americas. The manufacturing processes of individual industries, or even domestic habits within the home, have a direct and lasting effect upon the earth's environment. Even the most private behavior of individuals affects and is affected by large-scale social patterns in society as a whole.

If my marriage ends in divorce, it increases the likelihood that others' will do so, too. It destabilizes a whole nexus of wider social relationships.* If I cheat on my husband or in my business affairs, it increases the possibility that others will be less faithful or less honest. At the same time, the choices and decisions that others are making, the statistics on crime, mobility, abortion, divorce, church attendance, and so on all have a subtle, and sometimes a direct, influence on the choices and decisions I will make. With modern communications and computer data processing, I am made constantly aware of the trends in my society and the chances are increased that I will follow them.

Mechanistic notions that society consists of isolated units, each blindly pursuing its own self-interest, cannot cope with this interlinkage. Cobbled together out of separate, uncoordinated parts and yet impinged upon from every direction, mechanistic systems have become unwieldy and unstable. Old models of conflict and confrontation must give way to new structures for dynamic integration—structures that preserve the identities of participating members while drawing them into a larger working whole.

*There is a lengthy discussion of this in Chapter 13, "Reinventing the Family."

• *It must get beyond the individual/collective dichotomy.* We have gone as far as we can down the path of lonely self-discovery and private experience. The "culture of narcissism" described so powerfully by Christopher Lasch[19] is barren at its heart. It fails to nourish us either as individuals or as members of larger groups. At the same time, these past few years have seen the utter failure of collectivist attempts at social unification. I think not just of communism's dramatic collapse in Russia and Eastern Europe but also of widespread disenchantment with a more gentle socialism in Britain, Sweden, and other Western European nations; the decline of the kibbutz movement in Israel; and general loss of interest in the various commune movements popular in America during the 1960s and early '70s.

Neither extreme individualism nor extreme collectivism can satisfy our growing need to see ourselves as creative individuals within a larger and meaningful whole. We need to evolve a new alternative, a third way that mediates between the self-centeredness and fragmentation of extreme individualism and the imposed communality of extreme collectivism. Our whole sense of what it is to be a community must be transformed.

• *It must be plural.* The old vision of one truth, one expression of reality, one best way of doing things, the either/or of absolute, unambiguous choice, must give way to a more pluralistic vision that can accommodate the multiplicities and the diversities of our new experience. Learning to live with many points of view, many different ways of experiencing reality, is perhaps the greatest challenge of the new, complex society in which we find ourselves. Either/or must give way to both/and. "My way" must yield to a shared way that respects many possibilities as valid and many truths as steps along some further evolutionary way.

• *It must be responsive.* The new society must be flexible and less hierarchic. It is evolving. Ambiguity, rapid change, and greater complexity increasingly dominate events and patterns of relationship in both domestic and public life. Shifting boundaries of responsibility and identity (personal, cultural, national, sexual, and gender), experimental modes of living and family structure, new technological systems, new information sources, new and shifting markets all demand flexible response. Mechanistic patterns of fixed roles and rigidly organized structures for management and control fail to make full use of the vast potential latent in human response and imagination. We are not machines. Living systems were *designed* to cope with ambiguity and creative challenge; the future requires that we use this.

  • *It must be "bottom up," or emergent.* We have grown impatient with "top down" social and political structures, structures imposed by tradition, heredity, revelation, or outside authority. We no longer accept their authority without question. We can and should use the past as a lesson and a guide, but it can no longer be the final arbiter of conduct or belief. There is something deeply radical about the contemporary social and political consciousness. It is crying out to articulate a new vision that is emerging from within, from the grass roots where individual people think, feel, and act. We must relocate the source of authority and decision, make room for it among the frontline citizens, frontline workers, even frontline soldiers.*

  • *It must be "green."* To meet the awesome threat posed to the earth's environment by the waste products of industrial civilization, we need a social reality that is at one with the natural world around us. Both Christianity and mechanism assumed a dichotomy between human beings and the material world and between culture and nature. We must transcend this with a sense that human social goals can and should evolve in harmony with the wider living and geophysical context within which society is embedded.

  • *It must be spiritual.* If we are to experience deep satisfaction in our social lives, we must be able to see society in a larger context of meaning and value, a context that transcends the concerns of both materialism (or consumerism) and limiting self-interest. Our social vision must have a teleological dimension. That is, we must be able to answer questions like what is society *for*, what is its purpose and direction, in what dimensions of underlying reality do we find its roots, its systems of value, its moral foundations? These are, ultimately, spiritual questions. They have to do with how we understand the ultimate meaning

---

*My reference to authority issuing from frontline soldiers comes from NATO's new military philosophy known as "mission analysis directive control." It aims to get frontline officers accustomed to local analysis of the situation, local decision making, and "lateral thinking."

and sanction of our actions and projects. Such concerns were the motive force behind the founding of most religions, but spirituality itself is less organized than religion, less tied to any specific dogma or practice. A spiritual dimension in society need not (indeed, in a plural society, should not) be identified with any particular organization or group.

• *It must be in dialogue with science.* The old science and the traditional, organized religions have found themselves at loggerheads about the history and nature of the universe and the nature of human beings. At worst mechanistic science openly contradicts religious belief, and vice versa. At best they agree to differ and go their separate ways, each declaring the other irrelevant to its own interests. But advances in scientific knowledge make continuing apartheid increasingly unsatisfactory. With our loyalties divided between the vision of science and the claims of religion, we have no clear set of principles to guide our behavior.

If we are to rediscover the moral and spiritual roots of our society, *we must do so in a way which mirrors, which extends and develops rather than contradicts, the knowledge that science is giving us about the nature of the physical and living worlds of which we are a part.* Such dialogue makes no sense if the science we have in mind is mechanistic science but, as we shall see, the radically new science of the twentieth century is more compatible with our spiritual intuitions. Taking its insights on board may actually *help* us to articulate a more "modern" spiritual and moral vision.

Many of these qualities we hope again to find within our social reality were features of past, more traditional societies. For this reason, there is a strong sense of nostalgia linked to many of today's "progressive" social ideas. The Green Movement often harks back to more rural, preindustrial models for its economic and social ideals, just as postmodern architects and traditionalists long to rebuild the cities of the past and various New Age social thinkers seek a rediscovery of tribal society.

We should value greatly the past and its lessons. We must enter into a more creative dialogue with tradition. But in meeting today's social needs, the way back is not the way forward. "The problem," says Paul Ricoeur, "is not simply to repeat the past, but rather to take root in it in order to ceaselessly invent."[20] Some new dimension is needed. But to get from our still dominant mechanistic models of self and society, and from all the attitudes and patterns of behavior that go with them, to something genuinely new that has features like the vision I have outlined, will require a revolution in our perspective on social reality. Like all great social revolutions, this one will need a solid conceptual foundation if it is to succeed. It requires nothing less than that we adopt an entirely new philosophy of life, a whole new metaphysics.

During the past century, there has been a transformation of the same magnitude in our perception of physical reality. Today's physics requires that we learn to see the physical world in different terms. The old Newtonian categories are inadequate for the new understanding. Physicists have already learned that they are dealing with a new reality. Many features of the new physical reality mirror uncannily qualities we are now hoping to evoke in a new social reality.

The familiar certitudes of classical physics—rigid categories of space and time; solid, impenetrable matter; and strictly determined laws of motion—have given way to the strange world of quantum physics, an indeterminate world whose almost eerie laws mock the boundaries of space, time, and matter. A new kind of physical holism has replaced the classical emphasis on separate parts, and new patterns of dynamic relationship replace the old tension between isolation and collision. Where classical physics drew a sharp line between human beings and the material world, the creative dialogue between the observer and the observed in quantum physics suggests the possibility of an altogether more integrated relationship between ourselves and physical reality.

I shall argue throughout this book that we can draw on the principles and dynamics of quantum reality to derive a whole new model for our social and political relations. I believe this model can extend to a new vision of social reality that embraces every aspect of our daily lives—our sense of personal identity, our relationships to others and to nature, our political and moral decisions, the manner in which we design our cities and educate our children, the management practices with which we run our industries, and the fundamental values and goals that inspire our actions.

The project of defining such a quantum society is similar to that of the social and political thinkers who fashioned the modern age. They, too, often looked to physics to provide an overall conceptual structure for their ideas. I think they were right to do so, but they looked to the wrong physics—seventeenth-century physics is the wrong physics to meet today's social needs.

Nobel physicist Leo Esaki is perhaps the first to suggest that quantum physics provides a useful language for describing features of his own native Japanese society. In contrasting Western individualism with what he calls Japan's "collective mode," he says that Japanese society might most usefully be likened to a "coherent quantum state" or to a "room-temperature superfluid," a "many-body physical phenomenon in which the constituent individuals are modified in their behaviour and the whole is more than the simple sum of the parts."[21]

We will have to await our own later discussion of the kind of quantum phenomena represented by superfluids to understand fully the significance of his remarks, but he is attempting to use quantum analogies to describe the group cohesion and sense of common purpose that distinguishes Japanese society and industry. It is a cohesion that allows its members to "work together without friction," that gives the group an added identity over and above the identities of its individual members, and that provides the group with a creative edge not possessed by the individuals themselves. All these are characteristic properties of a superfluid (or any other very coherent quantum state). Anyone familiar with the practices of Japanese management will know that the "modification of individual behaviour" Esaki describes is brought about by getting the individual workers to function "in sync," or "on the same wavelength"—through a strict code of obedience, through group calisthenics, the communal singing of the company anthem each morning, the security of lifelong employment, the instilling of a high level of group pride, and a stress on the importance of group achievement.

As I shall argue in the pages that follow, the model of a quantum society I am attempting to develop for the West will stress the *joint* importance of what Esaki calls "the individual mode" and "the collective mode." (My own language for describing these is "the particle mode" and "the wave mode.") In this important respect a Western quantum society will differ in important ways from Japan's highly (and almost solely) group-oriented society, but it will also differ more radically still from the exaggerated individualism of our familiar, mechanistic Western models.

A great many sociologists, philosophers, and writers have, of course, expressed deep dissatisfaction with mechanistic science and the culture it has spawned. I think, for example, of men like Hegel, Max Weber, Jung, Sartre, Thomas Berry, and Peter Berger. There have been many more. But all too often these critics of mechanism then choose to ignore science altogether or remain ignorant of the radically new discoveries and altered perceptions made possible by the new science. They prefer, instead, an uneasy coexistence with science, very often postulating another, more human or more sacred realm that is somehow beyond science. Often this leaves their intuitions without the more solid foundation that science could provide. Sometimes it leads to an unresolved conflict between the claims of intuition and those of science, a conflict that weakens both. In the pages that follow I hope to show that there need be no such conflict where intuition is informed by the insights of physics.

This book will be written on two levels, the metaphorical and the speculative. At the one level I am content to argue that a quantum model of society is simply that, a model, an apt and powerful metaphor through which we can reorder our social and political thinking. Such a metaphor is justified by the many striking analogies that link the behavior of human groups and that of quantum systems, analogies that will become obvious as we go along. But these same analogies may imply that the idea of a quantum society has a more solid foundation still in the actual physics of human consciousness and human relationship. A "quantum society" may be more than just a model. It may be a very real possibility inherent in human consciousness.

It is my conviction that everything that is has a physics, and that therefore it is possible, at least in principle, to arrive one day at a genuine physics of society. At a second level, then, I will use the quantum model of how the brain gives rise to our conscious life to speculate how this social physics might really work.

Finally, whether we are speaking metaphorically or speculating about how society might actually function, I think it is peculiarly appropriate in our times to look for society's conceptual foundations amidst the discoveries of the new physics. Both socially and spiritually, we have a great need to see ourselves more within the context of the surrounding natural world.

Physics is not some separate and remote field of learning. Like everything else conceived by the human mind, physics is a product of consciousness and a product of the evolution of consciousness. Many of the big conceptual changes that have rocked physics in this century have mirrored similar changes in biology, politics, philosophy, art, and literature. There seems, as British novelist Nicholas Mosley has noted, "some impetus for change on every level," an impetus that suggests "there may be connections between these apparently different orders of things—the human and the scientific."[22] But in physics we find these changes highly articulated. Physics has a very precise language and clear set of categories for describing reality. It is also a *testable* vision of reality. Through the precise and testable imagery of physics we may learn a new language for describing other, related domains of our more daily experience.

There is another reason that physics may be a very appropriate conceptual model for our new social vision. That is the new global dimension to our lives, the fact that increasingly we are brought face to face with people from diverse cultures and traditions. We need to find our common roots in a conceptual structure we can all share, whatever our differences.

Physics is a universal language. The knowledge that it gives us, and the images and metaphors associated with this knowledge, are the common currency of every people on earth. As one of the founding fathers of quantum physics, Werner Heisenberg, noted,

> Modern physics is . . . a very characteristic part of a general historical process that tends towards the unification and widening of our present world. . . . Through its openness for all kinds of concepts it raises hope that in the final state of unification many different cultural traditions may live together and may combine different human endeavors into a new kind of balance between thought and deed, between activity and meditation.[23]

A "quantum society" seeks to achieve such balance.

# 2. Learning to Think the Impossible: Basic Elements of Quantum Reality

"I can't believe that!" said Alice. ". . . one can't believe impossible things."

"I daresay you haven't had much practice," said the Queen. "When I was your age, I always did it for half-an-hour a day. Why, sometimes I've believed as many as six impossible things before breakfast."

Lewis Carroll*

The social patterns and institutional structures that we evolve reflect the way that we structure our own ideas and experience in thought. Patterns of thinking hold us in their grip. They dominate the inner world of mind and constrain the possibilities available to us in the outer world of social reality. Sometimes they even evoke that reality itself. "There is nothing," as Hamlet reflected, "either good or bad but thinking makes it so."

If, then, we want to change society, we must begin by changing the way that we think. Ephemeral changes to our daily thoughts about what kind of society we would like to live in are inadequate for any deep social transformation. Thinking on this

*Through the Looking Glass.

37

level leads more often to "trendy" solutions for half-understood problems. Real social transformation requires that we change our basic *categories* of thought, that we alter the whole intellectual framework within which we couch our experience and our perceptions. We must, in effect, change our whole "mindset," learn a whole new language.

Our existing social perceptions and attitudes, the way that we think about ourselves, the way that we relate to each other and to the natural world, reflect our long immersion in mechanistic thought. The mechanistic categories of substance and identity, of space, time, movement, causality, and relationship are the categories that we use every day as we go about our personal and public business. We have grown so used to these old patterns of thinking that they seem the only "natural" ones to adopt. Most of us use them unwittingly, unaware of the extent to which they frame our lives. Faced with a suggestion that there is some other, some wholly different, way to perceive reality, we feel at a loss, perhaps even outraged. In the physical sciences, quantum physics has caused that kind of outrage.

Comparisons between features of quantum reality and the bizarre world of Lewis Carroll are common. The point is usually to stress the sheer impossibility of quantum physics and its supposed assault on all the accepted categories of common sense about the physical world. Many physicists themselves have echoed Alice's "I can't believe *that*!" when confronted with incredible phenomena like cats that are both alive and dead at the same time ("Schrödinger's cat") or with whimsical particles that change their identities for no apparent reason. But my own sympathies are with the White Queen.

I think that with a little practice we, too, could learn to believe six "impossible" things before breakfast—and that it could transform many of our everyday thinking processes if we were to try. I say this because I think the seemingly impossible nature of quantum reality has been overstressed, and its important lessons for our perception of both physical and social reality therefore overlooked. I think that the problem is not so much with quantum reality, but with what we take for common sense—with the conditioned nature of our own intuitions. As the Nobel physicist Richard Feynman put it, "The paradox is only a conflict between reality and your feeling of what reality ought to be."[1] I think we can learn to change what we expect from reality.

The "new" physics, as quantum theory is so often called, is nearly seventy years old. Some elements of it were discovered even earlier, at the turn of the century. In terms of its ability accurately to predict experimental results to several decimal places, it is our most successful physical theory ever. Its practical applications have given us nuclear power, laser beams, and the microchip, all of which have transformed our technology. Yet the details of quantum physics, and the sweeping conceptual revolution that underpins it, have made almost no impact on our perception of ourselves or the world around us. We simply don't understand it, and most of us, like Alice, think that we can't.

Those few nonphysicists who know anything about quantum physics think that it describes the behavior of very small things, the tiny microworld within the atom, a world that we can't see. That much is true, but it also describes large things like lasers, superconductors, and superfluids, and very enormous things like neutron stars and the interstellar laser beams (masers) that transmit coherent radio signals between the stars.[2] In recent years, physicists have discovered that the physics of life depends upon quantum phenomena in the cell walls of plants and cell membranes of animals[3] or even inside the DNA molecule itself.[4] In the next chapter I will argue that these same phenomena also give us the physics of consciousness.

In fact, quantum physics accurately describes *all* physical phenomena.\* The rough, classical physics that we are accustomed to using when talking about the everyday world about us is just a good approximation (an approximation that works well at certain restricted levels) of the real physics that quantum theory outlines.

So the problem is not that quantum reality is terribly remote or inaccessible. It is everywhere around us and inside us. In fact, as we get more in touch with the true nature of our own perceptions and conscious experience, I think we can appreciate the truth

---

\*Except the bridge between quantum reality and classical reality—"collapse of the wave function." This is a big gap in modern physics.

of David Bohm's claim that "quantum mechanics has been experienced by everybody far more than classical mechanics."[5] But we have been immersed in the mechanistic paradigm for so long that we have come to expect very particular things from the physical world. We are used to the Newtonian categories of space and time and solid matter and we can't easily imagine a world that mocks their existence.

Each time we walk from one room to another in our houses, we are aware of the distance between the rooms and the time it takes to go from one to another. We are aware of their separateness and of the solidity of the walls that divide them. We know that a hand pushing on a door causes it to open and gives us access to one of these separate rooms. Pushing and pulling against objects is the way that we force ourselves through our physical environment.

Our social perceptions, too, are colored by these categories of separateness, distance, and causality. We are conditioned to see ourselves as isolated islands of experience connected across the many separations that divide us by causal links arising from forces of "power" and "influence." The very way that we move our bodies and relate to the bodies of others illustrates this in bold outline. We carve out a space around ourselves, we hold ourselves back, we literally push against the space of others to keep them at bay. When we do reach out it is to push or to grab, to *force* others to move their bodies in some way amenable to us, to gain power over them or to defend ourselves against them. This characteristic Western "body language" is especially noticeable to anyone who has visited the Far East, where Buddhist culture has instilled a wholly different attitude to the nature of the physical world, including the body and our relation to the bodies of others.

We perceive our social institutions, our economies, and our nation-states as essentially separate and linked externally by these same chains of power and influence. In both business and international relations we often speak of "forcing" the situation. We move others, we manipulate them, we cause them to behave as we would like or we take measures to defend ourselves against their manipulations. We feel that we must have an effect on others or be affected by them.

When quantum physicists tell us there is no distance between objects, or indeed no solid objects in the sense that we mean them, and that the whole notion of "separate" has no basis in reality, we are at a loss. When they ask us to give up our commonsense notions of time and tell us that "causation" is not the only way for things to be connected, we are left not knowing how to structure the events and relationships around us. We are left with a choice between crying, "That's *impossible!*" and a realization that we must learn to see the world around us in a new way.

My purpose in this book is to make that new way of seeing seem more natural, indeed even quite attractive. I want to begin here by discussing some basic features of quantum reality—the way that quantum "things" exist, how they change and how they relate. As we go along, we can perhaps begin to see how the quantum realm reflects overlooked features of our own experience and how a renewed sense of that experience can reeducate exhausted and needlessly narrow habits of thought.

## *Being*

With our Newtonian way of looking at the physical world, we are very used to thinking that things have a very definite position or identity. Dozens of times each day I remind my young children that I can only be in one place or do one thing at a time. I am either at home or I am out, my sweater is either blue or it is some other color, either I am in a mood to go out or I feel like a quiet evening at home.

Physicists, too, have traditionally made such either/or distinctions. An electron was thought to have a very definite position and momentum. It was either within the range of a measuring apparatus or it was not. A chemical compound (such as benzene) was thought to have one possible electron ring configuration or another. For years classical physicists argued whether the basic constituents of light were really waves—vibrations in some underlying "jelly" (the ether)—or really particles—solid and separate "bits" each occupying a definite place in space and time.

Quantum physics has called this whole *either/or* way of thinking into question. When dealing with quantum reality, we have to learn a new *both/and* kind of thinking. We have to learn to get beyond apparent contradictions. For people schooled for centuries in the either/or mold this can be difficult. As management expert Richard Pascale says,

> It is almost impossible to think certain thoughts because our minds haven't been trained to work that way. We continue to polarize choice in terms of Theory X *or* Theory Y because we are wedded to a simpler intellectual framework. We can't readily conceive of things in and/also terms.[6]

I believe a greater literacy about quantum reality can help us with this.

One of the most revolutionary ideas thrown up by quantum reality is that light is *both* wavelike *and* particlelike *at the same time*. This is known as the "wave/particle duality." Neither aspect of the duality, neither the wavelike properties nor the particlelike properties, is more primary or more real. The two complement each other and both are necessary to any full description of what light *is* (the Principle of Complementarity)—and yet we are always condemned to see only one at a time.

We can design experiments in which light *behaves* like a series of waves, and others in which it *behaves* like a stream of particles. But we can never see its duality. According to one of the most fundamental and important principles of quantum theory, the Uncertainty Principle, we can never pin light down and say, "Reveal yourself as you really are." We can't measure all its properties exactly. If we treat it like a particle and measure its exact position, we will get a very fuzzy reading of its momentum (energy). If we treat it like a wave and measure its exact momentum, we will get an equally fuzzy reading of its position.

The problem of sizing up the exact identity of a photon (a bit of light) or an electron is somewhat like trying to give an accurate account of what it is like to participate in an excited crowd or in an unfamiliar ritual ceremony. If we give ourselves over to full participation, we know what it *feels* like to be a participant, we know the sense of belonging or the sense of being carried away, but we may have only a fuzzy picture of what we are actually a part *of*. We haven't stood back from the crowd or ritual and analyzed what is going on or who else is doing it with us. On the other hand, if we stand back from all the excitement like a journalist or commentator, we may be able to give a very detailed account of what the crowd is doing or what the ritual is actually about, but may have only the vaguest sense of what it must feel like to be part of it. As outside observers we are not caught up in the sense of the ritual or transformed by its possibly ecstatic dimension. A full experience of the ritual would require *both* participation *and* description, but that is impossible. Of course in familiar rituals, like those of the Catholic mass or the Passover ceremony, long experience of participation may combine the two points of view because we will, over time, have done some of each and wedded the two pictures.

This inherent uncertainty of quantum reality, its both/and character, replaces the familiar fixedness of the mechanistic world. Machines are very definite things, the same in all circumstances. Their performance may alter slightly in a given environment—they may rust if it is damp or seize up if contaminated by dirt—but they don't change *internally*. They don't become something else altogether.

An electron or a photon (or any other elementary "bit"), on the other hand, is in a constant creative dialogue with its environment, with the overall context of the whole experimental situation in which it is being measured. Like homonyms, words that look the same but have a different sense depending upon the context in which they are used (e.g., "lead," and "minute"), quantum reality shifts its nature according to its surroundings. In quantum philosophy, this is known as "contextualism."

The full force of quantum contextualism shows itself most dramatically in the famous "two-slit experiment." In this experiment, a stream of photons is emitted from a source. Just in front of the photon source, the experimenter erects a barrier with two open slits in it which can allow the photons to pass through. Beyond that, he places either two particle detectors (photomultiplier tubes) or a wave detector (a screen) with which he hopes to observe the photons as they strike. If he selects the particle detectors (measures the photons separately), the photons travel through *one*

## FIG. 2.1 TWO-SLIT EXPERIMENT

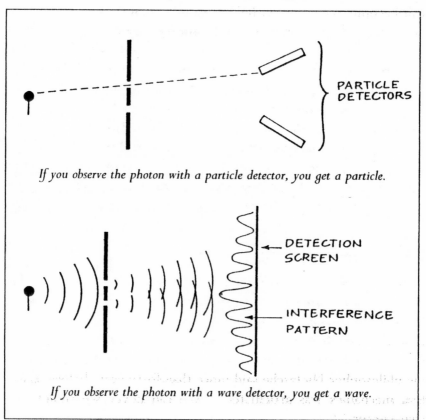

*If you observe the photon with a particle detector, you get a particle.*

*If you observe the photon with a wave detector, you get a wave.*

of the two slits and make a click in one of the detectors. If he selects a screen (measures the photons collectively), they travel through *both* slits and leave a wave interference pattern on the screen (Fig. 2.1).

*See F. A. Wolf.*

The two-slit experiment is often discussed to illustrate the creative relationship between the observer and the observed in quantum reality. Before the photons are observed (measured), there is no way that we can say they are really waves or really particles. They have the potentiality to be both. But when we observe them, when we erect either particle-detecting devices or a wave-detecting device in their path, the *type* of observation we use *evokes* one or the other of the underlying possibilities. What we see is what we look for. The overall context of the experimental situation, including the experimenter's own expectations, affects which side of its dual, underlying potential quantum light will become. The context helps reality to real-ize itself.

The relationship between the observer's way of looking at a quantum experiment and the outcome of what he sees is very like the link between our social expectations and what we perceive. If we look at a group of people as a collection of individuals, we will perceive them as individuals. But if we look at the same group as a collective unit, take an "average over the individuals," we will see a collective phenomenon. More strongly still, the way we look at a group of people can actually affect the group's behavior, or vice versa. Any public speaker or actor knows that the appreciation or the boredom of the audience has a strong effect on his or her performance, even if this is not expressed openly.

In dealing with quantum reality, then, we must always consider the whole situation. We can never, as we could in mechanism, isolate bits of the situation and analyze them independently. Neither can we, in sharp contrast with mechanism, isolate ourselves from the situation. In the quantum realm, the observer plays a crucial role in bringing about the very situation that he observes. His presence and his expectations physically alter what he sees. As the philosopher Nietzsche said more than forty years before quantum mechanics was articulated, "We can never see round our own corner."[7]

In the two-slit experiment, if the physicist *looks* for a particle (uses a particle detector), he will *find* a particle; if he looks for a wave (uses a screen), he will see a wave pattern before him. The physicist acts as a midwife to reality. Through his interventions, he helps to evoke one face of reality's rich, underlying potential.

In modern social philosophy, too, the importance of overall context and of the observer's "situation" have begun to figure in discussions of meaning, truth, and value. The French philosopher Merleau-Ponty, for instance, has said that when we speak of truth, we can only "define a truth within a situation."[8] Too often, however, these philosophers argue that truth and value have no foundation in reality, no existence *beyond* the context in which they appear. This has led to a dangerous social relativism that does not reflect the relation between reality and context in the quantum realm. In quantum physics, the "truth" that shows itself in any given situation is but one visible aspect of that situation's deeper, and very real, underlying possibility. In quantum reality, there is something *there,* beneath the manifestation.

The whole of quantum reality is in fact a vast sea of potential. That is the radical newness of the quantum realm. It is a realm of being where both/and is the rule. At the level of culture and society, there are clear analogies in the way that social circumstances evoke some particular side of our multifaceted human nature. All infants are born, for instance, with the potential to speak any one or several of the world's existing languages (they can and do utter in infancy all of the eight hundred or so phonemes which make up these languages), but each will develop his or her linguistic abilities in response to the language of the surrounding culture. We all know of children who are well behaved at school but naughty at home, or who are responsive to one teacher but dull with another. At the level of fundamental physical reality, the potential for such varied response is almost infinite.

In the quantum realm, the wave/particle duality and the creative dialogue between quantum potential and experimental circumstances shows us that there is always more to reality than we can experience or express at any one time. Adopted as a wider social paradigm, greater sensitivity to the latent potential of situations might encourage us to think about things not just as they *are,* but in terms of where they are going, what they will *become.* This could give us a more evolutionary outlook.

Viewing reality itself as so many patterns of shifting, responsive potential is alien to our mechanistic intuitions and to the rigid social structures and attitudes derived from them. We have grown used to experiencing things and people in terms of fixed identities and predictable patterns of behavior—to taking them "at face value." But it was not always so. Through the insights of quantum reality we find ourselves brought back full circle to our own more ancient perceptions and philosophical roots.

The early Greek philosophers taught that "all is flux" and that "nothing ever is, all is becoming."[9] In Aristotelian physics, *potentia,* or the capacity for becoming, was one of the primary properties possessed by a thing—the acorn "possesses" the property of possibly becoming an oak, the baby of possibly becoming a man.[10] In today's society, with its shifting boundaries of identity and responsibility and its constantly changing demands, a renewed sense of the creative potential latent in such flux might greatly enhance our own powers of response.

## *Quantum Transformation*

One of the most seductive features of classical physics is the underlying simplicity with which events are supposed to unfold. Newton's world is not static, but neither is it very surprising. One thing follows another in strictly determined order and with entirely predictable result.

If we know the starting position of a mechanical system, and the details of all its interactions on the way, then the mechanistic laws will tell us exactly where it is going and how it will get there. This is true even of the "chaotic" systems that excite so much current interest—they aren't really indeterminate, they are just astoundingly complex. Sometimes predicting their outcomes defeats even our best computers, but that is just because the computers are limited, not because the systems are inherently indeterminate. Things are very different in quantum reality. There, the indeterminacy is built into the reality. It is an *inherent feature* of the reality.

The same quality of true indeterminacy and multiplicity that permeates being in the quantum realm also characterizes the way that any change occurs, or the process by which one quantum system is transformed into another.

The word *quantum* itself refers to a discrete packet of energy, the smallest possible unit of energy that can be associated with any single subatomic event. In the early days of quantum theory, the physicist Max Planck proved that all energy is radiated in these tiny, somewhat "lumpy" units rather than smoothly and continuously as classical physics had thought. A few years later Niels Bohr demonstrated that electrons in atoms jump from one energy state (orbit) to another in discontinuous "quantum leaps," the size of the leap depending on how many quanta of energy they have absorbed or given off.

The old Bohr atom is now dated, but it is still useful for

**FIG. 2.2 THE BOHR ATOM, WITH A HEAVY NUCLEUS AT THE CENTER SURROUNDED BY ELECTRON ENERGY RINGS**

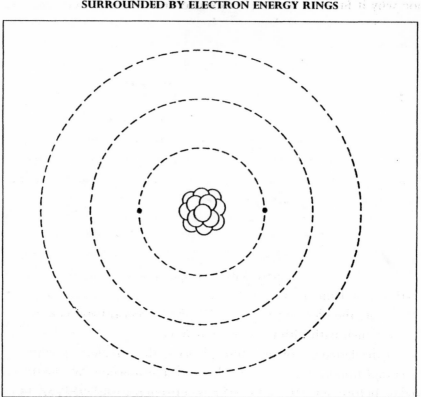

picturing the role of indeterminacy in the "leaps" that transform the quantum realm. Bohr pictured the atom like a minute solar system, with a heavy nucleus of particles at the center and rings of electron energy levels surrounding it. The individual electrons whiz around the nucleus in these rings, a bit like planets traveling around the sun (Fig. 2.2). In a stable atom, each of the electrons is "at home" in a particular orbit, depending on the energy with which its circling is associated. But strange things begin to happen when the atom becomes unstable—when its internal energy patterns begin to shift.

To begin with, the atom may become unstable for no apparent reason. There are no necessary "whys," or causes, for events in quantum reality. Things may just happen as they happen. So, quite suddenly, the electrons in a previously stable atom may begin to move into different energy orbits. And if they do so, there is no

*How can it choose. To choose means it has "mind." (?)*

way of knowing by which path a particular electron might travel, nor why it finally "chooses" to settle down in some other orbit. All that we can say is that its path will be in discontinuous, quantum jumps, and that the distance (energy difference) it travels will be measured in so many whole quanta, or "lumps" of energy.

Each possible journey and each eventual destination is associated with a probability, but nothing is ever determined. Indeterminacy—the lack of any physical basis for predicting the outcome of events—characterizes the quantum realm. The electron *may* go to the next lowest state, it *may* go to the next highest state, it *may* leap over several intermediate states or even double back on itself.

Worse still, from the point of view of accepted common sense, quantum physics tells us that the electron actually follows all these possible paths *all at the same time.* It behaves as though it is smeared out all over space and time and is everywhere at once,[11] In much the same way that we play with multiple possibilities in our imaginations, or launch "trial balloons" to see how something might work out, the electron puts out "feelers" toward its own future to see which path ultimately suits it best.

In the language of quantum physics, these feelers are known as *virtual transitions*—they are the possible journeys the electron makes before something actual (something measurable) happens. The actual journey, the one that results in the electron's finding a new home and staying there, is called a *real transition.* But the distinction between virtual and real is misleading and means nothing 'like it would in the world of common sense. As the physicist David Bohm warns,

> Sometimes permanent (i.e. energy conserving) transitions are called *real* transitions, to distinguish them from the so-called *virtual* transitions, which do not conserve energy and which must therefore reverse before they have gone too far. This terminology is unfortunate, because it implies that virtual transitions have no real effects. On the contrary, they are often of the greatest importance, for a great many physical processes are the result of these so-called *virtual* transitions.[12]

The relationship between the many possibilities of virtual transitions and the one actuality of a real transition is a bit like that between the plethora of opinion polls that precede an election and the actual election result itself. An opinion poll represents a *possible* reality, a way that people *might* vote if they could vote on that day.

Just as there are many possible and contradictory virtual transitions all happening at the same time, so we often see several opinion polls, each with a different supposed result, all published at the same time. And though an opinion poll is an election that never "really" happened, its results do have an effect in the real world. People often change their intended votes because of them. Government policies very often change, sometimes governments even collapse, in the wake of the polls' *possible* results.

The existence of virtual states shows us that we can experience more than one reality at a time, each playing out its individual drama simultaneously with that of others. In quantum language, these multiple realities are known as "superpositions." We get one reality literally "on top of" another. In the quantum realm, superpositions are the norm. The quantum wave function (or Schrödinger wave equation—the mathematical construct that describes any piece of quantum reality) always contains a plethora of possibilities, all equally real and many mutually contradictory.

Social reality, too, I believe, contains such a plethora of possibilities latent within it. And like their quantum equivalents, the social possibilities (lifestyles, cultures, languages, points of view, religious perspectives) are sometimes mutually contradictory. Perhaps if we could also discover some social equivalent of the wave function—the underlying common reality—we could find a unity in our differences, a positive grounding for our social pluralism. I will discuss this at length in Chapter 6 and later chapters.

To dramatize the utterly curious multireality of quantum superpositions, and their eventual "collapse" into one single, everyday actuality, physicist Erwin Schrödinger introduced his famous quantum cat.

Schrödinger's cat is kept inside an opaque box, beyond the range of all observation. With the cat in the box, there is also a fiendish (radioactively controlled) device that decides, randomly, whether to feed the cat healthy food or give it poison. Given our mechanistic, either/or logic, we would expect that if the cat is fed poison he will be dead, whereas if he is given food he will live. But this whole box is a small quantum world in which all things are possible. So long as he is not observed, the cat exists in a superposition—he is *both* alive *and* dead at the same time. It is only when we open the box to look at him (measure him, in physicists' parlance), that the cat's state must collapse into a choice. Peering into the box, we will find that he is *either* alive *or* dead (Fig. 2.3).

In quantum language, the moment of observation, the moment when many-possibility quantum reality condenses into a single actuality, is known as the "collapse of the wave function." Nobody yet understands for certain why wave functions collapse, only that it seems necessarily connected to their being observed (measured). In our example of the many simultaneous opinion polls, the moment of collapse is equivalent to our casting our actual ballots on the day of the real election.

When we think about it, many events in our everyday lives seem to be preceded by "feelers" toward the future followed by eventual "collapse." At the borders of consciousness itself we experience a plethora of fuzzy "prethoughts" that collapse into clear, single ideas when we concentrate. In our imaginations we constantly throw out, and experience, a multiplicity of future scenarios before these collapse into one at a moment of choice.

I think, for example, of what sometimes happens in social situations. If I go to a party and fall into conversation with an interesting but irritating man who feels the same about me, there is a range of possibilities. We may continue to talk, we may plan to meet again, or we may disagree and end our conversation. We are both aware of all these possibilities, and they affect the way we relate now. All are virtual realities. As sociologists Brigitte and Peter Berger say, "Society is constituted by the meanings of those who live in it."[13] The virtual realities are part of the meaning of our talking together.

On a larger scale, David Bohm has drawn a parallel between quantum virtual transitions and superpositions and the process of biological evolution.[14] We can think of evolution's trial runs, the mutations that don't survive, as Nature's way of exploring her many possibilities. And although, like the virtual transitions, many

## FIG. 2.3

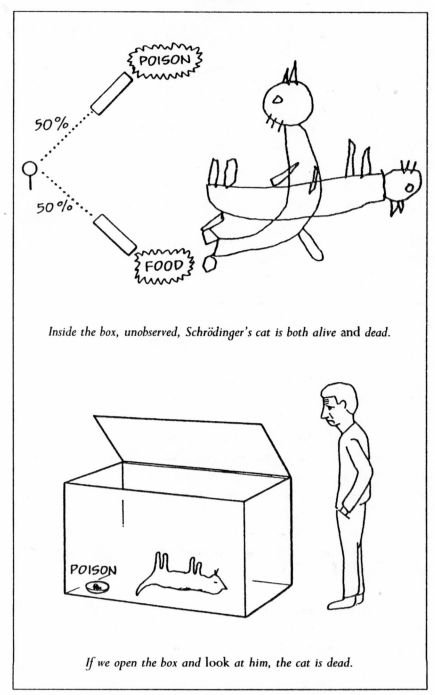

*Inside the box, unobserved, Schrödinger's cat is both alive* and *dead.*

*If we open the box and* look *at him, the cat is dead.*

of the trial runs themselves die out, they often leave traces of themselves in the new species to which they have given rise before fading away. We shall discuss this whole idea at great length in Chapter 7.

Many of our historical and cultural processes seem to follow these same patterns. I think, for example, of the hundreds of local Christian sects, many of them Gnostic, that preceded the rise of orthodox Christianity. All existed simultaneously. Any one of these *might* have "taken off," and several certainly did leave theological traces in the orthodox canon. But only the sect that led eventually to the Church of Rome did develop into a major and long-lasting cultural force, perhaps because this church had the qualities required by those times—or it might just have been luck (indeterminate).

On a much smaller, more individual scale, we often speak of "testing the ground" before making some social move, or of "putting out feelers" to see whether some decision is viable. While we do so, "all things are possible."

## How Quantum Systems Relate

In the world of classical physics, all of reality is ultimately reducible to basic, unanalyzable parts (atoms). Each part is inherently separate from every other part and connected to others only externally, through locally recognizable exchanges of force or energy. If one separate bit of reality moves, it does so because something definite has caused it to do so. When two separate bits meet, they collide and the force of impact sends them off in opposite directions.

The concept of relationship between any mechanistic entities is thus always one of an external connection mediated by some outside force or signal. Newtonian atoms can't get inside one another; they are impenetrable. They can't relate internally. In our social perceptions, these impenetrable atoms become the individual "units" of society whose necessarily external relations are mediated by power and influence, suspicion and mistrust.

In quantum physics, both the nature of being as a dynamic wave/particle dualism and the notion of transformation as a process through which things like electrons and photons are spread out all over space and time carry enormous implications for the kinds of relationships found between quantum systems. It is here, in the realm of relationship, that quantum reality is truly most "mind-boggling" and revolutionary.

Just as solid, Newtonian particles that meet must clash and go their separate ways, wave fronts that come together tend to overlap and combine. The reality of each is taken up and woven into the other. Quantum systems, with their potential to be both particles and waves, have a capacity to relate on both terms.

When two quantum systems meet, their particle aspects tend to stay somewhat separate and maintain shades of their original identities, while their wave aspects merge, giving rise to an entirely new system that enfolds the originals. The two systems relate internally, they get inside each other and evolve together. The new system to which their overlapping gives rise now has its own particle and wave aspect, and its own new corporate identity (Fig. 2.4). It is not reducible to the sum of its parts. We can't say, as in classical physics, that the new system is composed of *a* plus *b* plus the interactions between them. It is a new thing, an "emergent reality." In the physical world, such emergence is unique to quantum reality.

We can visualize quite simply what it means for two quantum systems to meet and preserve the individuality of their particle aspects while at the same time merging their wave aspects by doing a very simple thought experiment with two copper (or any other metallic) paper clips. Despite their size, such ordinary, household paper clips are everyday examples of quantum systems—it is impossible to describe the electric (and some other) properties of metals adequately in the old, classical physics. We can grasp the quantum nature of the paper clip by focusing on the atomic structure of its wire. Inside each paper clip, the copper atoms of the metal crystal are arranged so that the bulk of each atom (the ion— nucleus plus inner electrons) has particlelike properties, and the electrons in the outer rings have wavelike properties. If we run an electric current through the paper clip, the particlelike ions will stay in place, but the wavelike electrons will extend themselves along the wire of the clip, filling all its space and time (Fig. 2.5).

**FIG. 2.4**

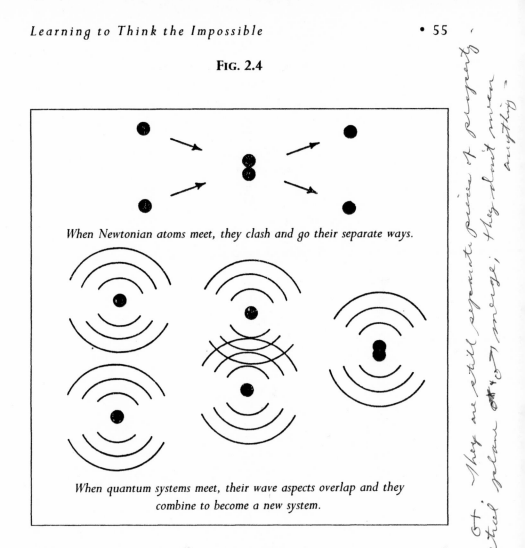

When Newtonian atoms meet, they clash and go their separate ways.

When quantum systems meet, their wave aspects overlap and they combine to become a new system.

This is why electric currents travel at half the speed of light—much faster than particlelike electrons could move through a metal.

Now, we can take two paper clips and join them by folding one into the other. I found it very useful actually to take two paper clips in my hand and join them while thinking about this experiment. When the clips are joined, the particlelike properties stay separate and maintain their original identities. We are still left with two solid, discrete paper clips, each with its own distinct boundary. But the wavelike electrons making up the two clips' electric currents will *merge*. They will become *one* electric current

*[handwritten marginal note, right margin:] Like σ + σ, They are still separate pieces of property. On the actual plane σ+σ merge, they don't mean anything.*

*[handwritten note, bottom:] non-gender true reality.*

## FIG. 2.5 EACH PARTICLE HAS A PARTICLELIKE AND A WAVELIKE ASPECT.

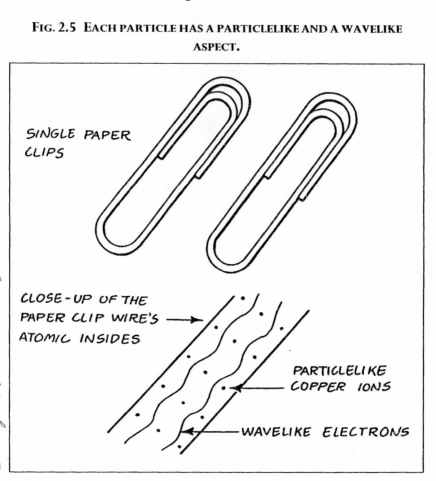

SINGLE PAPER CLIPS

CLOSE-UP OF THE PAPER CLIP WIRE'S → ATOMIC INSIDES

PARTICLELIKE COPPER IONS

WAVELIKE ELECTRONS

with a very slightly modified wavelength. So overall the two clips folded together represent a new quantum system with particlelike properties similar to constituent systems' but with entirely new wavelike properties* (Fig. 2.6).

Human beings who experience intimate relationships know something of this meeting from the inside and the way that one's

---

*The effect is increased dramatically in superconductors and superfluids, where the whole system is in a more wavelike state to begin with because there are fewer irregularities (due to thermal "noise") among the constituent particles. In the same way, following through on Leo Esaki's earlier analogy between superfluidity and the coherent nature of Japanese society, Japanese individuals are "modified" to cultivate group values and to behave as members of a group. Therefore we might expect Japanese society to be more wavelike (more like a superfluid) than Western society, where differences between individuals are emphasized and cultivated.

A good example of this in with man & woman. On Earth in the 3-D you have separate gender, marriage. On the 4-D or higher realms, you have non-gender merging.

**FIG. 2.6 THE PAPER CLIPS' PARTICLELIKE IONS STAY SEPARATE, AND HENCE THEIR SURFACES REMAIN SEPARATE AS WELL, BUT THEIR WAVELIKE ELECTRONS MERGE.**

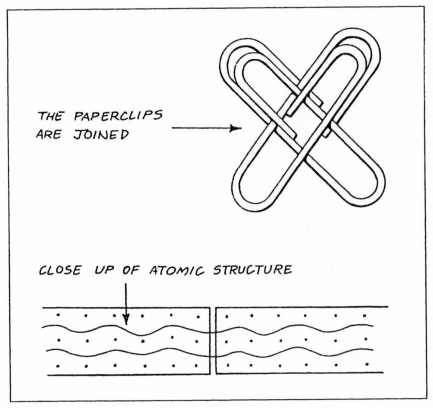

individual identity can be both preserved and at the same time taken up into a new whole that is somehow larger than the self on its own. I believe the same dynamics apply at the larger social level, and give us our first inkling of how a society might be more than just the sum of its individual parts and their external relationships. I shall discuss such emergent social reality, and its role in providing the cement that binds societies together, when we come to discuss the emergence of persons and community.

Overlapping quantum systems are one form of internal relationship we find in quantum reality, but they are by no means the most dramatic. It is when we consider the implications of things like electrons and photons being smeared out all over space and time that we discover a kind of definitive relationship that robs the word *separate* of its usual sense and makes us rethink completely what we mean by parts and wholes.

For if all potential "things" are spread out infinitely in every direction, how do we speak of any distance between them, or conceive of any separateness? If all things and all moments touch each other at every point, the oneness of the overall system is of a kind hitherto unimagined. In describing such extreme interconnectedness, quantum physicists have revived the once ghostly notion of "action at a distance," in which one thing can be correlated with another instantaneously with no apparent exchange of force or signal between them. Known as "nonlocality," or correlation in the absence of any local forces, this somewhat eerie interconnectedness is one of the greatest conceptual challenges thrown up by quantum reality.

It was Einstein who first showed that the equations of quantum mechanics implied a kind of instantaneous connectedness between apparently separate things. This seemed to raise problems for the most basic principle of relativity theory, which holds that nothing (such as a signal) can travel faster than the speed of light. Thus relativity tells us that there can be no such thing as an instantaneous *causal influence*. But without "causes" and "influences," Einstein argued, nonlocality was "ghostly and absurd" and its prediction was clear proof that quantum theory was wrongheaded or incomplete. He tried to prove this by suggesting a paradox—the famous Einstein-Podolsky-Rosen, or EPR, Paradox.

The gist of the EPR Paradox can be illustrated with an example of two twins. Suppose that I, living in Oxford, have an identical twin who lives in New York. Imagine us to have been separated in early childhood and to have had no contact over the years. Yet in many observable ways, our lives seem to be strangely correlated. I live in a three-story house and have two children, so does she. I am married to a doctor, so is she. My favorite color is red, she dresses in nothing else. Each of us earns her living writing popular books about quantum mechanics, and so on. How can all these correlations be explained?

A quantum physicist would find the question easy. He would say his equations have always predicted such correlation effects, and that they show the twins' apparently separate lives are in fact

*I call this the voice of God hollering at itself.*

both aspects of some larger whole that naturally keeps them in synchrony. For Einstein, this made no sense. In his Theory of Hidden Variables, he said that there must be some mysterious common factor which accounted for all the similarities. In the analogy of the twins, this might be our shared genetic material. The controversy wore on for years, but was eventually settled by Irish physicist John Bell. His theorem—Bell's Theorem—led to experiments that finally settled the matter.

If we applied Bell's Theorem to my twin and myself, we would look for some instantaneous correlation in our lives that couldn't possibly be accounted for by genetics. Imagine, for example, that I decide to enroll in a dance class in Oxford. My twin, too, might be observed to enroll in such a class in New York, at exactly the same time. Strange, but perhaps not absolutely outside some genetic predisposition linked with coincidence.

But supposing that I am observed to raise one or the other of my arms during the exercises. What does my twin do? If quantum physics is right, and important aspects of our lives are indeed instantaneously correlated, she should raise one of her arms (the opposite one) at exactly the same moment. If Einstein is right, my raising an arm should make no difference to what she does—no shared genetic material can account for simultaneous arm raising, and no faster-than-light signal can travel from Oxford to New York to tell my twin what I have done.

In fact, if we are quantum twins, every time I am seen to raise my right arm, my twin raises her left one. If I raise my left arm, she raises her right one, at exactly the same moment. Which arm I raise at any time is completely indeterminate—both my twin and I are fully ambidextrous. Each of us chooses freely which arm to raise in the dance. And yet our behavior is linked as though we are standing in the same room and there are springs connecting our arms. Instantaneous, nonlocal correlations may be "ghostly and absurd," but they are a fact of quantum reality.

In social reality, too, there often seems to be an eerie correlation between apparently separate events or situations—the spontaneous appearance of revolutions or the acceptance of new practices all over the globe, or the almost simultaneous discovery of creative ideas by two or more people working in ignorance of each other. In the theater, director Peter Brook has spoken of an almost "telepathic intuition" that establishes real contact between a group of actors "who have no common language or references, no shared jokes or grumbles." The actors just "get into sync."[15]

Sociologists and historians describe these phenomena as "trends" or "coincidences," and they may have all manner of explanations. But it is just possible that something like nonlocal quantum correlation effects may operate at the social level. The possibility becomes more likely if there is indeed a quantum-mechanical basis to human consciousness. *"Likes attract!"*

In the quantum laboratory, the twins of my example are actually represented by pairs of photons that have been "introduced" to each other and then fired off into different directions. When the polarization of one is measured, the polarization of the other is found, instantaneously, to be exactly the opposite. They are always negatively correlated.[16] The two photons could be on opposite sides of the room or, theoretically, on opposite sides of the universe. They are so eerily linked that, whatever their apparent separation, they behave as though there is no space between them. They are parts of the same larger whole and this whole seems to choreograph their simultaneous movements, like some mysterious web of connection wrapping them both in its influence.

These same quantum correlation phenomena have also been observed across time. Two events that appear to have happened at different times in fact unfold as though they were happening at the same time. It is as though both were present simultaneously.

Imagine, for example, two boys, each of whom has a peg for his coat at school. When both boys are at school, both pegs are occupied. But suppose one boy attends school only in the mornings and the other only in the afternoons. If these are quantum boys, whose lives are linked across time, we will find that the place where each hangs his coat is correlated with that of the other. So if the boy who comes in the morning hangs his coat on Peg A, the boy who arrives in the afternoon will always choose Peg B. If the first boy chooses Peg B, the second will choose Peg A. Which choice the first boy makes is completely indeterminate (because all quantum boys' choices are indeterminate), and he leaves no

*[margin notes:]* See "The Theatre of Renault" by Robert Brustein

certain Miranda and enchantress are the same.

St. Francis + Padre Pio → and Jesus.

**FIG. 2.7 THE TWO PHOTONS ARE FIRED SEPARATELY, WITH A TIME DELAY BETWEEN THEM. YET THE INTERFERENCE PATTERN INDICATES BOTH HAVE GONE THROUGH THE SLITS SIMULTANEOUSLY.**

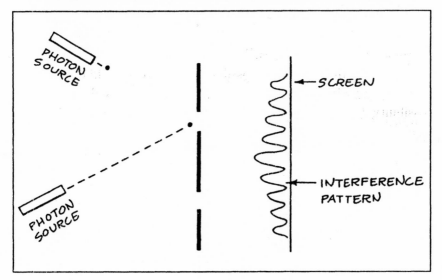

trace of his choice as a signal for the second boy. It is just that each boy will always behave as though both were present.

These across-time correlation experiments have in fact been done with laser beams. In one, two separate photons, from two separate lasers, have been fired through one of two slits in a barrier at different times. Though only one photon can strike a detection screen at any one time (our kind of time), the experimenter observes an interference pattern on the screen indicating that *both are present simultaneously*[17] (Fig. 2.7). That is, the wavelike pattern indicating the presence of the earlier photon crisscrosses with the wavelike pattern of the later one. This is impossible unless the earlier photon had ''reached across'' time to ''be there'' simultaneously with the later photon.

In a more recent experiment, one single photon is fired at a mirror covered with a very thin coat of silver. Because the coating is so thin, the photon has a 50 percent chance of passing straight through the mirror as though it were plain glass and a 50 percent chance of being reflected off it. If it goes straight through, the photon follows a short, direct path to the distant screen. If it is re-

FIG. 2.8 THE PHOTON HAS A 50 PERCENT POSSIBILITY OF TRAVELING THROUGH THE MIRROR AND ARRIVING MORE QUICKLY, AND A 50 PERCENT POSSIBILITY OF TAKING THE LONGER ROUTE. YET THE INTERFERENCE PATTERN INDICATES BOTH POSSIBILITIES HAVE ARRIVED SIMULTANEOUSLY.

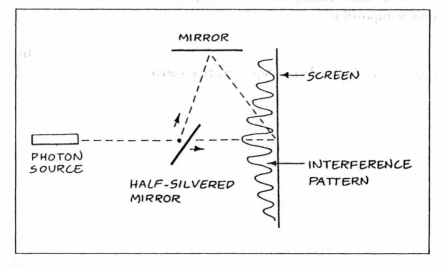

flected off, it follows a longer, roundabout route and reaches the screen later. It should be either/or. What we actually see on the screen, however, is an interference pattern—indicating that the photon has interfered with (crisscrossed with, or become enmeshed with) its own *possibility* (its virtual state) of arriving sooner or later[18] (Fig. 2.8). The mere possibility of the photon's arriving earlier or later has a kind of temporal reality that "smears" itself out across time.

The whole of quantum reality is to some extent an unbroken web of overlapping or correlated internal relationship. It has what David Bohm calls a quality of "undivided wholeness."[19] This is why quantum physics is said to be "holistic," but it is important to realize this is not the weak holism that has been used too much as a catchword for the New Age.

Quantum holism is not the supposed holism of a carpet with its separate strands woven into a piece, nor even the apparent holism of the earth with its many biosystems all functioning as a dynamic living unit.* In each of these cases, had we sufficient powers of analysis we could predict the new whole from the fixed properties of its parts and a knowledge of the interactions between them. We can always reduce the system to those parts and their interactions.

In the truly emergent holism of quantum reality, such reduction is impossible. Through this strong holism the new whole possesses qualities (and an identity) of its own that arise only *through the relationship* of its previously undefined (indeterminate) parts. In quantum reality, relationship is truly creative.

There are many parallels, I believe, between the role of indeterminism in the emergence of holistic quantum systems and important features of both human language and creative forces in social organization. I think that both may possess genuinely holistic properties, and that with our organizations, at least, we can enhance these through the right sort of structures.

Quantum holism may be telling us, for example, that power relations are not the only, or perhaps even the most effective, way that people and events can be linked in society. The politician or the manager who tries to ''influence'' or ''control'' events may be less effective than one who can be sensitive to the spontaneous emergence of social or political ''trends.'' The individual who realizes that parts of his or her own identity emerge through relationship with others may be less guarded and defensive.

The crucial importance of indeterminism for allowing the kinds of correlations that lead to true quantum holism may recommend looser organizational structures. Social or managerial organizations that constrict relations between people or groups with tight rules and rigid role definitions may be stifling creativity and the emergence of new ideas.

Throughout this book I will be drawing parallels between aspects of quantum reality and qualities we might expect to find or hope to evoke in human nature and community. But at this point, it becomes necessary to ask why there *should* be any parallel. What has quantum physics to do with *us,* or with our

---

*I am not speaking here about *Gaia*, which, if it exists, would be a truly emergent, holistic phenomenon.

social relations and organizational structures? Are the kinds of social correlations we have discussed *analogous* to quantum correlations, or are they *literally* quantum correlations? In the next chapter I shall begin to answer these questions by looking at how quantum processes may be playing an active role in the workings of the brain and how this could affect the emergence and the structure of consciousness.

I am convinced that in my retiring years my studying and reading all this stuff is a "homework gateway" to the next realm.

# 3. A New Physics of the Mind

*Perhaps it is not too fanciful to suggest that quantum correlations could be playing an operative role over large regions of the brain. . . . . It seems to me a definite possibility that such things could be playing a role in conscious thought modes.*

Roger Penrose*

We have seen that quantum reality offers us the model for a new kind of thinking and thus, perhaps, for some new social vision. But can we think of this in literal terms, as the way that thinking *really* works? Might there be quantum structures in the brain that grant us access to the insights of the quantum realm, and if so can we build these into a wider model of the human mind and of the social structures that our minds create? I think that we can, and that such a model can give us a broader appreciation of our potential for both thought and relationship.

There has been heated debate in the second half of this century about the relation between mind and brain. The great increase in our knowledge about the brain and its capacities has led to an assumption by most people that our thinking processes, our capacity for language, and perhaps even our social structures and institutions, must in some way be derived from, and constrained by, the physical nature of the brain itself.[1] At the same time, our ex-

---

*"Minds, Machines, and Mathematics."

isting models of how the brain works are mechanistic ones. They are derived from the principles and categories of Newtonian physics, and liken the thought processes of human beings to those of our personal computers. Such models reinforce the old, mechanistic patterns of thought. They fail to account for many of the qualities that we recognize and value about our experience.

At the most basic level, experience includes things like being aware, the capacity to feel pleasure and pain, the capacity to move about at will, and the capacity for spontaneous and purposive activity. In these simple ways, our own experience is probably much like that of at least the higher animals. In human beings, experience also includes other things, things like understanding, rationality, and imagination; a sense of meaning, of belonging and of loving. It includes emotional, aesthetic, and spiritual experience, an experience of self, or "I-ness," and of the mortality of that self.

All these qualities of experience belong to the realm of mind that we call consciousness. So far, no machine (computer) has been built that has consciousness and no mechanistic model of the brain can account for its existence. Machines (of the Newtonian variety) are made of separate interacting parts related by causal laws. Their behavior is always reducible to the fixed behavior of these constituent parts. Machines can imitate the *behavior* associated with conscious experience to some extent, but we cannot imagine them actually having the experience itself. We cannot imagine a machine experiencing something like toothache, a sense of humor, or an "identity crisis"—there is no one there, no emergent "personality," to act as the subject of the experience. This is just as true of machines made out of neurons as of those composed of microchips.

The problem of accounting for conscious experience on the basis of known facts about the physical brain has led to two diametrically opposed models of mind. Each has had consequences for the way we see ourselves as individuals and the way we live together in society.

On the one hand, there are those who still claim that the conscious mind simply cannot arise from the physical brain, that mind is its own distinct reality and must have some extraphysical origin. These people—Nobel laureate John Eccles and the philosopher Karl Popper are prominent among them[2]—are in the old dualist tradition of Plato, Descartes, and the Christian Church. They are not arguing necessarily that mind has a special spiritual dimension like a "soul," just that mind, whatever it is, is not physical. Their thinking encourages a split between mind and the physical world and a consequent split between nature and culture. It supports the alienation of human society from its wider natural context and underlies many of our current environmental crises.

At the opposite extreme are those who claim that all aspects of mind must have a physical basis in the brain much as we currently understand it. If our knowledge of the brain cannot account for conscious experience, they say, so much the worse for that experience. Perhaps consciousness is just an illusion, or some tenacious projection of an inner psychological need that satisfies biologically determined evolutionary requirements. But, even if consciousness were just an illusion, the question of *who* is experiencing this illusion would still raise obvious problems. Others then go on to argue that even if consciousness does exist, it safely can be ignored because it has no effect on behavior.* *PUKE!*

All these people—and most advocates of AI (artificial intelligence) are among them—would reduce mind either to those qualities that arise solely from mechanistic structures ("strong AI") or at least to those qualities that can be *simulated* by computing machines ("weak AI"). This is like saying that we capture the full essence of an orchestra when we hear its sounds simulated on CD. Strong AI (mind is nothing but a computer program) has been opposed passionately by some philosophers, notably John Searle,[3] and even weak AI has its critics (Hubert Dreyfus[4] and Roger Penrose[5] foremost among them) who argue that mind has further qualities that no machine possibly could simulate. Nonetheless, the AI machine-models of mind have had a marked effect on our models of thinking and organization, and have limited both to structures that emphasize logic, rule-following, rationality, and either/or distinctions. From these we build a society that is poorer than our actual capacity for experience and creative thought.

It seems to me obvious that the full range of conscious expe-

---

*Philosophically, this last group are known as "epiphenomenalists."

rience is both real and important. It is also most likely that this experience, like all aspects of mind, arises from some physical activity in the brain. But to argue both requires not only a deeper understanding of the quality and physics of mind but also a wider model of brain structure itself. As American philosopher Jerry Fodor sums up the current state of research based on mechanistic models of this structure,

> We know a little about language, a little about perception, very little about cognitive development, practically nothing at all about thought, and, as far as I can tell, nothing at all about consciousness. The problems about consciousness, in particular, have proved intractable in a very unsettling sort of way.[6]

Fodor, and others of his eminent colleagues, like philosophers Thomas Nagel[7] and Colin McGinn,[8] reach the pessimistic conclusion that we may never achieve a physical understanding of consciousness—that it is a problem outside our scope. But I think the alleged problem arises solely from the insistently mechanistic approach of today's cognitive science. If a brain structured on mechanistic principles cannot account for the rich world of consciousness, then the most fruitful way forward would be to look for a model based on other physical principles that can. I believe we can find support for this in the quantum realm.

## Does the Brain Contain a Quantum System?

It has been nearly half a century since David Bohm first drew attention to several striking parallels between features of our own experience and the way that quantum systems behave.[9] We have seen some of these in the last chapter.

Both quantum reality, for instance, and some features of both human language and human nature are "situational," or context dependent. They reveal different "sides" of themselves in different circumstances. Think, for example, of my saying to my husband when he walks into the room after dressing for a party, "You have your own unique style of dress." This might be a compliment, it might be a joke or a criticism, or it might just be a neutral description of his appearance. Exactly what kind of statement it is depends on how I feel, how I say it, the expression on my face, and so on. In the same way, we all know that our feelings and behavior may vary considerably depending upon where we are or with whom we are. Most of us feel and behave differently when on holiday in some foreign country than when at home. I am more spontaneous and charismatic when lecturing to an audience that appreciates my work than when speaking to one that is hostile.

Then, too, both indeterminate quantum systems and our human imagination involve "superpositions." They contain several possible realities all juxtaposed, one on top of the other, that they explore simultaneously by throwing out "feelers" toward the future. Quantum systems do this to test out the most stable future energy state; our imagination does it to test out the best possible future life scenario.

There is also the parallel between the constraints on a physicist's dialogue with quantum reality and the kinds of choices we must make. In his measurement, or observation, of a quantum system, a physicist must choose between an exact description of the system's position and an exact description of its momentum (the Uncertainty Principle). He can never know both at once.

In our own experience we find we must often choose between getting the "feel" of a situation or giving an exact report of its factual detail. In our thought processes we must choose between entertaining a vague train of thought and concentrating on one clearly focused idea. In our organizational structures we frequently must choose between the efficiency of a system (which is often linked to tight structure) and its potential for creativity (often requiring loose structure). There seems to be a kind of Uncertainty Principle operating here, too.

A further suggestive link between the experience of human selves and the behavior of quantum systems is the whole question of free will. We do experience ourselves as free and spontaneous beings. It seems to us that we can and do make free decisions, that we can and should bear responsibility for our actions. Yet if the mind works like a machine, there is no basis for this felt sense of freedom and responsibility in the actual physics of the brain.

Machines are wholly determined things. Each part of a machine

*[handwritten marginalia, left margin]* Oh, God, there we go again with that one!

*[handwritten marginalia, right margin]* This is at the heart of the genre of painting — different styles of painting in art. Abstract Expressionism vs the Uncertainty Principle. Traditional art is exact in position?

works as it must according to the forces acting upon it. Each element in my word processor works according to the program I insert onto its disk. But quantum physical systems are wholly indeterminate in the way they respond to their environments. Quantum indeterminacy acting through a conscious system in the brain would not in itself wholly explain the complex nature of free will and moral responsibility, but it would provide a physical basis *compatible* with such a human potential. No mechanistic basis to mind ever could.

Finally and, for our purposes, perhaps the most striking parallel between quantum reality and our conscious thought processes, there is the unity, or holism, that distinguishes both. Our thoughts "hang together" with what Roger Penrose has referred to recently as a "oneness" or a "globality."[10] They cannot be broken down into distinct elements that on their own make any sense. Penrose uses the example of the totality of a musical composition to illustrate this.[11] Other, more daily examples are numerous. I cannot, for instance, separate the peculiar beauty of my leaking and asymmetrical canoe from the fact that I built it with my own hands, nor the sweet thrill of drinking hot chocolate from childhood memories of my grandmother serving it to me.

This same holism is characteristic of my conscious field itself. As I sit here at my desk, my brain is bombarded at every moment by millions of sensory data—visual, auditory, tactile, and olfactory. And yet I do not perceive a scene fragmented into millions of parts. I perceive a room, I am aware of myself sitting in this room and of the projects that draw me here. My consciousness creates a unity from the diverse bits of sensory information, drawing them into a meaningful whole.

The unity of consciousness has always been the major stumbling block to finding any physical explanation for our mental life based on known brain structure. In modern neurobiology it is known as "the binding problem"—quite literally, how does the brain bind its disparate neural activities into an experience of a perceptual whole? This problem was the driving force behind Descartes' original separation of mind and body. "There is a great difference between mind and body," he said, "inasmuch as the body is by its very nature always divisible, while the mind is utterly

### FIG. 3.1 THE BRAIN AS A COMPLEX TELEPHONE EXCHANGE

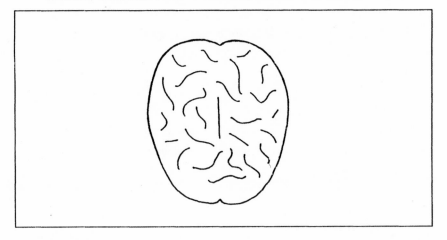

indivisible.''[12] All mechanistic structures are inherently divisible, and mechanistic models of the mind share this same quality.

In most standard brain models, our mental life is thought to arise somehow from signals passing back and forth along the brain's $10^{11}$ neurons, like electrical pulses inside a giant computer or communications through an enormously complex telephone exchange. The problem is that nowhere in such a mechanism is there any means for unifying the billions of neural events triggered by the constant bombardment of sensory data. There is no one neuron, or system of neurons, that acts like a railway signalman or air traffic controller to ensure that the mechanism works as a whole (Fig. 3.1). Neural networks are thought to integrate information within various brain ''modules,'' but this still leaves the problem of what integrates the modules. There is no way to see how ''I,'' the self, or the perceived wholeness of my world can emerge from a system consisting of so many separate parts. From whence comes the agent who initiates actions or who exercises will?

Socially, we have the same problem with models that view society as consisting of so many separate individuals. What, in such societies, is supposed to bind all these individuals together into a social unity? What is the imagined basis of social cohesion? Existing models based on a balance of conflicting self-interests or a balance of envy or ''social norms'' seem inadequate.[13] This problem is one of the great unanswered questions of modern sociology and political theory, and one to which I shall return at length in later discussion.

Many quantum processes, by contrast, have the same inherent indivisibility that we note about our mental processes. As with the elements of a poem, it can be impossible to separate any one element of a quantum event from the overall context in which it happens.* We can remember, for example, that in the famous two-slit experiment (Chapter 2) it depends entirely upon whether we measure a photon with a photomultiplier tube or with an interference screen that it manifests itself as a particle or a wave. The overall experimental situation decides the photon's identity and even writes its history. If it becomes a particle, we know that it will have traveled through one slit. If it becomes a wave, we know its journey will have taken it through two slits. This context dependence occurs because, at the level of holistic quantum reality, relationship itself *creates identity*—the identity (characteristics) of the whole quantum system emerges from the way its "parts" combine. Like the unity of consciousness, quantum holism cannot meaningfully be analyzed, or broken down, into those original separate parts.

Both Bohm and Penrose (and many others) have wondered whether so much analogy between quantum reality and features of our thinking might imply some actual physical connection between the quantum realm and human consciousness. There is by now a whole literature on the subject.[14] "Might there," asks Penrose, "be any relation between 'a state of awareness' and a highly coherent quantum state in the brain? . . . It is tempting to believe so."[15]

There is no doubt that quantum-level events happen in the brain all the time. The retina of the eye is sensitive enough to register the arrival of a single photon of light. The indeterminate firing thresholds at neuron junctions are most probably due to quantum fluctuations in the chemical concentrations at those junctions. There are similar quantum fluctuations in the activities of ion channels within the neurons themselves. Penrose has speculated that nonlocal quantum correlations may be playing a role in

*This is true of quantum processes that have got into a highly correlated state.

the growth and contraction of dendritic spines (neuron endings) at the synapse junctions between neurons in the brain.[16] The cause of such spontaneous growth and contraction is one of the big unsolved problems associated with human learning and concept formation.

It is indeed tempting to wonder whether features of our conscious life that cannot be explained by mechanistic physics might originate among similar quantum processes in the brain. I think particularly of the experienced unity of the self, or "I." No mechanistic system consisting of separate, interacting parts ($10^{11}$ neurons in our case) could give rise to the indivisible unity of our sense of self, but an emergent, holistic quantum substructure possibly could. That is, a structure that is greater than the sum of its parts, that has some identity over and beyond the identities of those parts, could account for how something like self could emerge and evolve over time. Our capacity for the free play of imagination and the unity of our conscious field are also qualities that *could* arise from a quantum physical system in the brain. In later chapters I will argue the same about our creativity and our ability to form new concepts.

Over the past several years there has been extensive discussion of a "holographic model" of the mind. This model originates with suggestions made by David Bohm and the California neurosurgeon Karl Pribram that there are interesting similarities between the holism, or unity, of mental images and the kind of holism found in holograms.[17] The exciting thing for advocates of this model is the way in which *each* part of a holographic image contains information spread across the whole pattern. That is, if we break the hologram apart into small pieces, we will still be able to get the whole picture from each of the pieces—just as each cell in the human body contains the genetic code for the whole body.

> In other words, each individual part of the picture contains the whole picture in condensed form. The part is in the whole and the whole in each part—a type of unity-in-diversity and diversity-in-unity. The key point is simply that the *part* has access to the whole.[18]

For Bohm, Pribram, and others, suggestions that mental images are similar to holographic images seem to illustrate how consciousness might be unified, and do so in a way that runs counter to the whole tone of mechanistic mind. The emphasis is on the "holism" of the model.

I think that holographic models are very promising, and worth pursuing, but in their present form they don't go far enough. The main thing, to my mind the crucial thing, lacking in existing holographic theories is that they don't discuss the underlying physical basis of holograms, nor do they attempt to discuss *how* a flesh-and-blood brain could function as one. Without this physics and this "how," the theories must remain a poetic model, or a metaphor.

Holograms are in fact the products of quantum processes. They arise from information "written" on a laser beam. In the language of physics, they are "excitations" of a laser beam. I want to discuss for a moment what it is about a laser beam that makes it such a special kind of quantum structure and then explore the possibility that we might find this kind of structure in the brain. If we could do so, we would do more than just validate something like the holographic model. We would have gone a long way toward establishing an actual quantum physical basis for conscious mind.

Laser beams are one reasonably familiar example of a quantum structure known as a "Bose-Einstein condensate" (so called because its properties were first suggested by Einstein and the Indian physicist Satyendranath Bose). A few other Bose-Einstein condensates are superfluids, superconductors, and neutron stars.[19]

The essence of any Bose-Einstein condensate is that it is, so far as we know, the most highly ordered and highly unified structure possible in nature. Its many "parts" are so unified that they "get inside" each other (their wave fronts overlap). They share an identity, or become as one whole. The many photons in a laser beam overlap their boundaries and behave as one, single photon. A physicist can write a single equation to describe the whole system.

It is also the case that Bose-Einstein condensates exhibit the highest degree of "agency," or causal responsibility, in the microphysical world. They seem actually to make "choices" about the possibilities open to them.[20] This may seem an outrageous thought, but as one prominent physicist expresses it,

> These ideas only seem strange when we apply them to micro-objects because, by habit and by virtue of our cultural environment, we think of micro-objects as inanimate. It is no doubt true that to assert that micro-objects have volition is too strong a statement, but . . . they have attributes that are the vague beginnings of volition and activity.[21]

These properties of Bose-Einstein condensates, their specifically quantum kind of unity* and their quantum "agency" could, in theory, provide a physical basis both for unified consciousness and for what we experience as free will.[22] They might also bear on our quest for more unified social structures. There is nothing like them in the old physics.

In classical, mechanistic physics, the most ordered structure that we can find is a crystal. Each atom or molecule in a crystal occupies a set position within a tight lattice structure, and all crystals of a given compound are identical (except for size). But crystals are hard, immovable structures. Their constituent parts always remain separate and they interact causally with each other via forces (of chemical bonding).

In the brain, crystalline structure is analogous to modeling mind on the ordered, causal interaction of separate neurons. In society, it is analogous to seeing separate individuals filling identical spaces in fixed social patterns, their behavior and relationships bound by rigid role-playing or social codes. Such societies (Victorian England, Prussian Germany) are highly ordered, but their citizens may have no sense of genuine social unity.

By contrast, classical physics also offers us the example of systems that have a high degree of unity, but no ordered structure. In a gas, for instance, the many molecules share a space. They overlap, and each can be identified only as a molecule of that gas

---

*Specifically quantum because only quantum "bits" can get inside each other, or relate internally. The constituent particles of classical physics are always inherently separate. They can only relate externally.

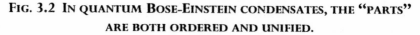

**FIG. 3.2  IN QUANTUM BOSE-EINSTEIN CONDENSATES, THE "PARTS" ARE BOTH ORDERED AND UNIFIED.**

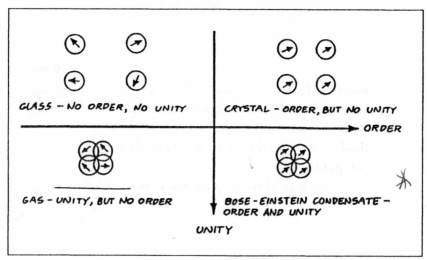

(they share an identity), but they relate to each other chaotically. Socially, I am reminded of family breakfasts with our young children—there is a great deal of unity (family relationship), but no order whatever.

In a Bose-Einstein condensate, however, the quantum properties of the structure allow both a "fluid" (ever changing) order and a high degree of unity* (Fig. 3.2). Each indeterminate particle in a Bose-Einstein condensate fills *all* the space and *all* the time in whatever container holds the condensate. Many of their individual characteristics are correlated—they behave holistically, as one. So the condensate as a whole acts as *one single particle*. There is no "noise" or interference between separate parts. This is why superfluids and superconductors have their special frictionless qualities, and why laser beams are so coherent.

If it were possible to get Bose-Einstein condensation (superconductivity, or laserlike activity) in the brain, it might indeed draw fragmented information from thousands of excited neurons

---

*The order arises from nonlocal (acausal) relations between indeterminate elements in the original parts; the unity arises from the ability of some quantum particles (bosons) to overlap and combine in such a way that they share an identity.

into a coherent and meaningful whole.* Such a structure could provide the physical basis necessary for the unity of consciousness—consciousness (and hence, the self, the "I" of conscious life) would literally *emerge* from the coherent ordering of the diverse bits of contributing information. It would have its own physical reality that is not reducible to the sum of its parts. All this is an ideal quantum basis for conscious mind, but even those who have suggested it[23] have found an insuperable problem—the temperature of the brain.

Superfluids, superconductors, and laser beams are all either very low-temperature or very high-energy phenomena. Brains are neither. Indeed biological tissue as a whole is neither. So all good theory aside, the question has remained: How could anything so warm and "sticky" as a brain support coherent quantum phenomena? The feasibility of a quantum physical basis to mind depends upon a positive answer to this question. It seems to me there is a very good candidate for one in the technical literature.

## *Body-Temperature Quantum Phenomena*

Within the past twenty-five years, physicists and biophysicists working in several different countries have been doing research on biological phenomena (hence, warm and "sticky" phenomena) that appear to be of a quantum nature. They have been working independently, but most take their lead from the pioneering work of Professor Herbert Fröhlich at England's Liverpool University. In the late 1960s, Fröhlich—himself an eminent solid-state physicist—predicted the likelihood of body-temperature quantum phenomena (Bose-Einstein condensation) in living tissue.[24] Experimental tests of his theory have verified the presence of coherent quantum structures in yeast cells,[25] bacteria, and inside the DNA molecule.[26,27]

In Germany, physicist Fritz Popp has discovered that biological tissue emits a weak "glow" when stimulated at the right energy

---

*The latest calculations on warm, laserlike, coherent quantum phenomena in water suggest a range of effect exceeding ½ mm. In the brain, this would extend across 10,000 neurons. More far-reaching effects are theoretically possible (E. Del Giudice, 1988).

levels. He sees this as evidence of photon emission from a coherent biophoton field, and suggests such quantum phenomena play a crucial role in cell regulation.[28] Working independently, Japanese scientists have found the same effects, concluding that they "are clearly associated with a variety of activities and biological processes."[29]

Fröhlich believed that the body-temperature quantum phenomena he predicted might account for biological coherence, the fact that living systems behave holistically, or have this elusive quality we call "vitality." In my earlier book, *The Quantum Self*, I suggested that Fröhlich-like phenomena could provide the physical basis for a quantum-mechanical model of consciousness. Recent experimental evidence in the neurosciences, combined with recent experiments on the link between quantum phenomena and consciousness, now indicate this is not likely to be the whole story. These "Fröhlich systems" themselves have characteristics* that make them unsuitable candidates, on their own, for the physical system underlying consciousness. Their effect is more likely felt at the level of coherence in individual cells. But Fröhlich's work does nonetheless have important implications for *any* attempt to describe consciousness in quantum terms. The experimentally verified presence of such "warm" quantum structures in biological tissue at all suggests the *feasibility* that other, more suitable quantum structures might be associated with neural activity in the brain. And since any other biological quantum structures might be expected to rest on the same basic dynamics, we can get a better concrete sense of what form these others might take by looking at a Fröhlich system in some detail for a moment.

Fröhlich-style quantum activity is concentrated in the cell walls (or membranes) of biological tissue. The wall of each cell contains countless protein and fat molecules, each of which may carry a tiny electric charge such that it is positive at one end and negative

---

*Fröhlich-style quantum systems are a microwave phenomenon, and they oscillate at a frequency of millions of cycles/sec. EEG (electroencephalogram) wave patterns associated with various states of consciousness oscillate at much lower frequencies, under 100 cycles/sec. A quantum model of consciousness would rest, ideally, on a physical system that oscillates at similarly low frequencies. This follows from the experiment done by C.M.H. Nunn, et al. (1992), discussed in a moment.

at the other. Because of their electric charge, the molecules are called "electric dipoles." When the cell is at rest (not stimulated, or being fed, by any energy input), the dipoles are out of phase, arranged in a haphazard way. When energy is pumped through the cell (energy derived from the digestion of food), these tiny molecular dipoles begin to oscillate, or jiggle, more intensely. And when they do oscillate, each "broadcasts" a tiny microwave signal, just as though it were a small radio transmitter.

What Fröhlich was to show was that when the energy flowing through the cell reaches a certain critical level, all the cell wall molecular dipoles begin to "line up," or come into phase. They begin to oscillate in unison, as though they are suddenly coordinated. And when they do oscillate in unison, the microwave that each generates independently suddenly gets pulled into one single coherent quantum microwave field. This emergent (it literally rises up from the oscillating dipoles) coherent field has the holistic properties common to any quantum field. This field is a Bose-Einstein condensate* (Fig. 3.3).

Such Fröhlich systems may well exist in the brain's neurological tissue, but whether they do or not, they provide an important clue to the kind of quantum activity that might arise in some other way within the brain. Given the most recent (since 1989) discoveries of neurobiologists working on the brain, it is even possible to speculate what brain dynamics might be associated with such activity.

Neurobiologists working on the binding problem (the unity of visual experience) have found that all the neurons in the cerebral cortex associated with any given perceptual object, say a cup, oscillate (vibrate) in unison. Thus any neuron, regardless of its location in the brain, that fires in response to my seeing a particular cup oscillates in unison with all other neurons that respond to that cup. And these coherent oscillations are the key to how so many disparate, and distant, neurons can integrate their information. Neurons oscillating in unison (at the same frequency) are a bit like amateur radio hams managing to communicate with one an-

*There is a much more lengthy discussion, in lay terms, of Fröhlich's work in *The Quantum Self*, Chapter 5.

## FIG. 3.3 FRÖHLICH-STYLE QUANTUM COHERENCE

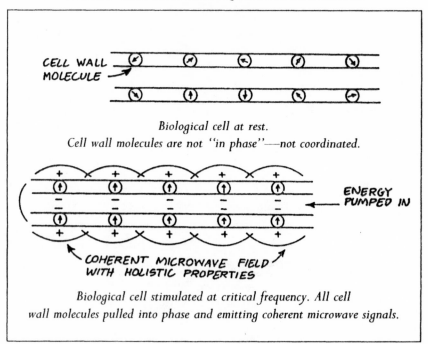

*Biological cell at rest.*
*Cell wall molecules are not "in phase"—not coordinated.*

*Biological cell stimulated at critical frequency. All cell*
*wall molecules pulled into phase and emitting coherent microwave signals.*

other by tuning into the same wave band. Coherent oscillations in the cortex are also associated with other senses, and with action, emotion, and memory. Research constantly extends the list.[30,31,32,33] This much is now neurobiological fact. But it remains a mystery *how* these coherent oscillations form—that mystery *is* the "binding problem." I think there is some reasonable ground for quantum speculation here. It may be significant that coherent oscillations also play a key role in Fröhlich's work on quantum structures in biological tissue.

As some molecular biologists have noted more recently, "Experimentally we know that there are several phenomena in which quantum effects are relevant to biomolecular dynamics."[34] Within the brain, the place to look for such quantum activity would be the ion channels (protein molecules) lining the membranes of individual neurons. Recent research supports this possibility.[35] These channels, which open or close in response to electrical fluctuations resulting from stimulation (sensory, motor, thinking, etc.), act like gates to let sodium, potassium, or other ions

**FIG. 3.4**

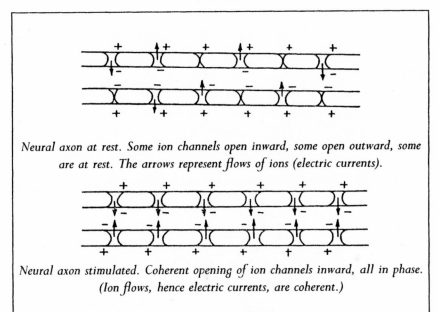

*Neural axon at rest. Some ion channels open inward, some open outward, some are at rest. The arrows represent flows of ions (electric currents).*

*Neural axon stimulated. Coherent opening of ion channels inward, all in phase. (Ion flows, hence electric currents, are coherent.)*

through. They are of a size to be subject to quantum fluctuation and superpositions. And their opening and closing can take the form of oscillations. Each ion channel, as it oscillates, generates a tiny electric field. When a large number of the ion channels (there are ten million in each neuron) open and close in unison, as they tend to do when stimulated, the whole neuron fires, or oscillates, and a larger-scale electric field is generated across the neuron. At a larger scale still, it is the collective electric field from tens of thousands of such coherently oscillating neurons that is measured by an EEG machine attached to the scalp (Fig. 3.4).

Certain neurons in the brain act as "pacemakers."[36] When these pacemaker neurons oscillate in response to stimulation (by coherent ion-channel openings and closings), whole bundles (modules) of neurons oscillate with them. This is like a tuning fork that makes neighboring tuning forks vibrate at the same frequency when it is struck. It is a resonance phenomenon. All the neuron bundles associated with the cup I see oscillate in unison. Other objects that I see are perceived through other neuron bundles oscillating in unison at different frequencies.

All this is orthodox, well-established neurobiological fact. My quantum speculation is simply to suggest that the original ion-channel oscillations are a quantum phenomenon, as in the Fröhlich systems, and that the consequent coherent electric field generated by the whole neuron is also a coherent, quantum electric field (a Bose-Einstein condensate). That, too, would be analogous to Fröhlich's coherent microwave field. By a steady step-up process, it would then be possible to imagine that the large-scale coherent electric fields known to exist across the brain are themselves quantum in nature.

## Why Must Mind Have a Quantum Dimension?

There are three reasons to suspect that the brain's coherent oscillations have a quantum basis. One is the unity of conscious experience and the fact that, among physical systems that we know, only a Bose-Einstein condensate possesses that kind of physical unity. A second reason is the speed with which the brain forms its coherent oscillations. It has to decide which of the myriad sensory data go together to form distinct objects, and these objects themselves may be multicolored or partly overlap. For a visual event, the brain does all this sorting within one-tenth of a second. If the brain were a standard serial, or even a parallel, computer, one neurobiologist has calculated it would need more time than the age of the universe to perform all the necessary calculations associated with just one perceptual event.[37] But if the brain were a quantum computer, it could try out all the various *possible* combinations of data arrangement at once,* and thus unify its experience quickly. This is exactly in line with Penrose's discussions about how quasicrystals can form so rapidly. These involve the nonlocal assembly of atoms in a compatible way. Penrose[38] speculates that all the feasible arrangements form in a quantum superposition, then collapse into a lowest-energy arrangement.

The third, and quite likely the most substantive and exciting, reason for suspecting a quantum basis to the brain's coherent os-

---

*Because of quantum superpositions, where an evolving quantum system tries out all its possible futures at once. See Chapter 2.

cillations is experimental. In 1992, a team of physicists working at Southampton University in England designed an experiment to test whether the coherent electrical fields known to exist across the brain and associated with various thought processes are quantum. These are the EEG patterns. To do this test the team attached EEG electrodes to both the right and left hemispheres of several dozen subjects' brains. The subjects didn't know whether the electrodes were switched on or off. Their goal was to see whether switching the EEG electrodes on had any impact on conscious task performance. If the brain's electric fields are nonquantum, measurement should have no effect. But if the electric fields associated with thinking are quantum, measurement should "collapse" them—i.e., noticeably alter focused thought and hence task performance.

In one run of tests, the numbers zero through nine were flashed on a screen at the rate of one per second. Each subject had to press a button with his or her right thumb if the digits flashed were one of a preselected three of these (two, five, and eight). Since both numeracy and right-thumb movements are controlled by the brain's left hemisphere, this was predominately a "left-brain" task. Performance on this task was *improved* by measuring EEG waves on the brain's left side, but noticeably disrupted by any measurement on the right side. Presumably this was because measurement on the right side caused the subjects to focus on something other than the task—perhaps, for instance, on the background music that was being played to mask any sounds of the measuring machine itself.

The outcome of the team's first run of tests was definitively positive. The subjects' performance was altered by measurement to an extent that was calculated as being one thousand to one against chance. The same team has since done a second run of experiments in which subjects and experimenters were in different rooms to exclude any tacit signaling. They got the same dramatically positive results. An EEG machine measuring the left hemisphere noticeably improved the accuracy of tasks carried out by the subjects' right hands.[39]

This is a very significant experiment, the first one I know of that directly tests for a quantum basis to conscious brain activity.

It is currently being repeated by other research teams. If others get the same positive results, it all but proves there is a quantum dimension to mind.

Full-blown consciousness itself, the rich world that we experience at the everyday level of full awareness, is far more complex than the simple unifying properties of a coherent quantum field, however it is generated. If the brain's known coherent oscillations are producing such an emergent (electric) field,* it is simply the "blackboard" on which our experience can be written. Or, with a better analogy still, we might think of the background state of consciousness (the Bose-Einstein condensate) as a pond, and all the contents of consciousness—our thoughts, images, emotions, memories, etc.—as ripples on that pond.

If we remember back to the holographic model of mind, the brain's Bose-Einstein condensate would play the role of the laser in that model, and "ripples" on that condensate would produce thoughts and perceptions akin to the hologram itself. A hologram is a "ripple" on, or a modulation of, the laser light's underlying uniform field. Such modulation generates the information that then gets represented in the holographic image. Similarly, the mind's thoughts and perceptions would be information "written" on the brain's condensate. By contrast, we know that in a state commonly produced by various forms of meditation, the mind seems to return to a state of pure, background consciousness. This is associated with a widespread, coherent electrical rhythm (EEG waves)—i.e., a pond without ripples or a laser without a hologram superimposed.[40]

A physicist would call the ripples that make patterns on a Bose-Einstein condensate "excitations," meaning patterns on the "surface" of the condensate. Such patterns would be generated by electrochemical activity arising from the brain's synaptic activity and triggered by the vast array of inputs that go to make up a person—our genetic code, our experiences, our instincts and emotions, our memories and bodily needs. It would be the role

---

*The oscillations certainly do generate an electric field (hence EEG patterns). The question is simply whether this field has quantum characteristics, i.e., whether it is a Bose-Einstein condensate.

of the Bose-Einstein condensate to draw all these various inputs into a coherent and meaningful whole. Each of us would experience this whole as ''I,'' the self that we feel ourselves to be, and the world that ''I'' inhabits.

At the social level, the emergence of a group or national ''collective self'' (the ''British people,'' the ''American people'') from the disparate inputs of the various individuals, their experiences and habits, their activities, fears, and aspirations might be analogous. I will discuss this much more fully further on.

## *Two Complementary Brain Systems*

By adding a quantum dimension to the physics of the mind, we end up with a model of brain function that has at least two interacting systems.

At the one level, we would recognize and appreciate the validity of all that conventional, mechanistic cognitive scientists have proposed about neural pathways (the brain's ''wiring'') and their role in the brain's capacity for information processing. There can be no question that such neural activity is involved in sifting and analyzing the constant stream of sensory data with which the brain is confronted. At this level, we probably do function like very complex computers. But any computer model of the mind treats a neuron like a two-state switch—it is either ''on'' (firing) or ''off'' (not firing), like a computer element. This is more or less true of the ''output'' end of the neuron—its axon—but it is not true of the ''input'' end, the dendrites. It is these dendrites that give rise to the slower, graded oscillations in ion channels that I have been discussing.

*The implication of this is that classical neuroscience, which can be modeled on computers, is only telling us half the story.* It, literally, is discussing only one end of the neuron.

So I am suggesting that there is a second ''level'' or system in the brain, working in tandem with the computational system (which itself embraces both digital and parallel processing). This second system is very likely some sort of quantum system. We do at the very least have a complex network of coherently oscillating neurons. It would be from this system that we could hope to gain

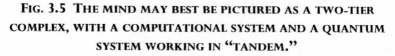

FIG. 3.5 THE MIND MAY BEST BE PICTURED AS A TWO-TIER
COMPLEX, WITH A COMPUTATIONAL SYSTEM AND A QUANTUM
SYSTEM WORKING IN "TANDEM."

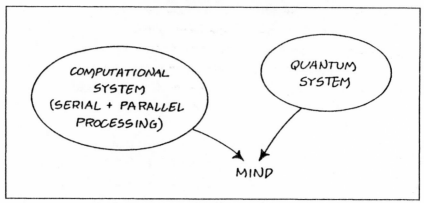

our capacity to *make something* of all the information available to
us—our ability to integrate it into the meaningful whole that is
the unity of consciousness, or the unity of the self (Fig. 3.5). The
two systems, classical and quantum, would not be anatomically
separate. Both would function simultaneously all over the brain,
and both would be necessary to account for our characteristically
human intelligence. Together, they give us important insight into
the nature of human thinking and its embodiment in social reality.

Given the two known classical types of processing done by the
brain—serial and parallel—plus the suggested quantum type of
processing, the brain may then embody *three* kinds of interacting
thinking or processing systems. Serial processing, as in our personal computers, manipulates the symbols of a language according
to fixed rules. The associated kind of thinking corresponds quite
well to the rational Freudian Ego or to the operation of a bureaucracy.

The more recently discovered parallel processing[41] works by
association, not reason. It can build up complex habits and recognition skills, such as riding a bicycle or learning to recognize
faces, but has no "language" to describe its operations. This whole
associative level of learning is prelinguistic and can be found in
"lower" animals. The kind of thinking associated with this resem-

bles that also attributed to the Freudian Id, that part of the psyche associated with instinct and association rather than with reason and language, which are functions of the Ego. Psychotherapists try to disentangle these sometimes pathological associations by bringing them into the sphere of logic and language. When it is healthy, associative thinking (or parallel processing) offers quite a good model for the subtle but unreflective and inarticulate behavior of traditional small groups. (See discussion in Chapter 11.)

Both serial and parallel thinking are classical (prequantum). Neither can show creativity or transcend the limits of its built-in structure—its rules or program. By contrast, a kind of quantum thinking produced by the brain could exhibit holistic, emergent features. As I shall discuss in later chapters, this facility might bear on our art, morality, and relationships and may help us to deal with the creative challenge posed by our increasingly pluralistic society.

## *Quantum Features of Conscious Experience*

Together with all the analogical evidence for some physical link between quantum physics and consciousness, this three-tier model of the mind and associated types of thinking might help to explain some features of our mental (and, we will see later, our social) life that hitherto have remained mysterious.

There is, for instance, the curious relationship between brain damage and consciousness. When various regions of the brain suffer damage, we experience a loss of function associated with those areas. Thus damage to the optic cortex can lead to blindness, damage to the motor cortex to paralysis, damage to Broca's area to loss of speech, and so on. But people suffering even very severe brain damage often remain fully conscious. This is also true of lower animals and babies, where there is little, or as yet undeveloped, cerebral activity.

When brain function is solely dependent on neural connections, these facts are difficult to explain. If we interrupt the signal transmissions along a series of telephone cables at any point, whole sections of the system break down. So a mechanistic model of consciousness arising from such connections would move us to expect a similar disruption when neural damage occurs. But if consciousness arises from a coherent quantum field over large regions of the brain, a field in which neural information is evenly distributed throughout, then the capacity to remain conscious despite damage to (or a lack of) cerebral activity should remain intact.

In fact, we know that consciousness itself is lost usually only during sleep (when there is very little energy being pumped through the brain, and hence through the neural ion channels), after massive damage to the brain or a sharp blow to the head, or under the influence of anaesthetics. This last cause, the impact of anaesthetics on consciousness, may in fact prove a fruitful approach to studying the physical basis of consciousness in the brain and its possible links to quantum activity.

Schizophrenia, too, may be explicable as a disease in which the coherence of consciousness itself is disturbed by chaotic surges of energy (chaotic electrochemical activity) interrupting the smooth "surface" of the brain's otherwise coherent quantum field (or, more accurately, interrupting the establishment of coherent quantum activity across the brain). Schizophrenics show no deficiency of thought, indeed many are highly intelligent, but they have difficulty drawing their thoughts into an ordered and meaningful whole. Their conscious field is often quite literally fragmented, causing them to see parts of their own experience as alien—hearing voices, seeing visions.[42] In dreaming, we all experience a similar fragmentation, creating whole dramas from dissociated parts of our personalities. This "normal" fragmentation may result from the sleeping brain's lower energy levels, making less energy available to maintain the coherence of the conscious field (the Bose-Einstein condensate). If thought itself arises from neural information processing (neural connections), as the proponents of AI claim, but the ordered unity, and hence the meaning, of thought from quantum coherence in the brain, this clinical picture makes sense.

Various aspects of our sense of time have also remained mysterious. In classical physics there is no direction to time—all processes are time reversible (at the human scale)\* with the important exception of entropy. And yet our consciousness is structured to give us a clear, forward sense of flowing time. If our consciousness has a quantum mechanical basis, this may be linked to the time irreversibility built into quantum physics by the continual collapse of the indeterminate Schrödinger wave function—the collapse from many possibilities into one reality.[43]

We also have the perceptual characteristic that psychologist William James termed the "specious present"—the time interval over which we can integrate perceived events into a single experience. The specious present lasts for anything up to twelve seconds. It is the period we identify as "now" in our sense of flowing time. Events falling within it are experienced as a unity. Significantly, I think, this time duration is in the same range as the "coherence time" of a Bose-Einstein condensate (such as a laser beam)—the length of time over which the condensate can maintain its basic pattern, i.e. its phase relationships.[44] In a quantum model of mind, it is the phase relationships (between the oscillating neuron bundles) that give rise to consciousness.

A correlation between the duration of our perceptual "now" and the coherence time of a Bose-Einstein condensate is not a *sufficient* reason to suppose consciousness has a quantum-mechanical dimension, but it is a *necessary* condition. Were the two different, there would be no physical basis for supposing a link.

But more important still for all our future discussions about the nature and dynamics of society is the mystery of "I," the thinking subject, and how this "I" emerges from the physical activity of myriad brain parts.

Existing mechanistic models of mind cannot account for how a thinking subject could result from the inherently separate functionings of neural structure. Mechanistic models of society have the same problem accounting for how any social cohesion could emerge from the coming together of a myriad of separate individ-

---

\*There are irreversible physical processes at the cosmic scale (the expansion of the universe) and at the subnuclear scale (some particle decays), but neither of these can have much to do with human brain function.

uals or separate ethnic groups. The existence in the brain of an emergent, integrating Bose-Einstein condensate might give us a physical basis for both—for the unity of the conscious "I" and, perhaps, for social cohesion. If so, it might indicate how we can make progress with the urgent problem of how to sustain social cohesion in our increasingly pluralistic societies. In the following pages I will turn the quantum model to a discussion of these problems.

# 4. Private Selves and Public Persons: A Quantum Model of Community

*Persons are not quite the same as solitary individuals, nor are they a crowd. Persons are living networks of biology and emotions and memories and relationships. Each is unique, but none can flourish alone. Each in some way contains others, and is contained by others, without his or her personal truth ever being wholly isolated or exhausted.*

Angela Tilby*

"Persons," according to philosophers, sociologists, and moral theologians, are necessarily social and moral beings. Persons are what we become in and through participating in the common life of this planet.

Each of us can recognize something of ourselves in Angela Tilby's moving definition of persons. Each of us experiences ourselves as an individual with our own personal truth, our own unique style, set of emotions, and very individual conscience. Yet at the same time we feel that we only truly know ourselves, only truly *become* ourselves, through the complex set of relations that bind us to nature, to others with whom we are in daily commerce, and to the

*Let There Be Light, p. 8.

culture of which we are a part. Our individuality, we feel, can never be wholly exhausted but neither, we recognize, can it ever be wholly isolated. *Yes, even for monks,*

This personhood, or individuality within relationship, is one of the defining qualities of our human nature. Yet it is perhaps the most difficult to account for in any workable theory of self or community. At both levels, at the private level of the self and at the more public level of community, we have the problem of how to account for the possibility of diversity within unity, or the problem of the individual within the group. This problem has become more urgent with the complex demands made on self, the increasing pluralism of modern society, and the threat of fragmentation to both.

At the everyday, unreflective level of my experience, I take myself for granted. I behave as though I exist, I think of myself as an effective person capable of following my own will, and I rely upon the unity of my experience. "I" am a simple fact of my everyday life. And yet if, as a scientist or psychologist, I ask myself "Who am I?", or "Where do I come from?", the answer is not so simple.

As a scientist I am aware that my brain contains $10^{11}$ different neurons, each responding in its own way to the countless sensory stimuli by which I am bombarded at every moment. How does "I," my unified sense of self, emerge from such a cacophonous jumble? As a psychologist I know that my self is "a house with many rooms."[1] "I" am a whole community of often contradictory subselves and "complexes," each with its conflicting sets of memories and emotions, conflicting motivations and desires. How despite it all do these add up to "me"?

Again, as I live my everyday life I experience myself in relationship to others around me. Without reflecting I say that "we" like this or "we" will do that. This "we"—my family, my friends, my cultural group, or my nation—seems natural to me, as real to me as the unity of my self. But if I am a social scientist or a philosopher, this "we" raises questions I cannot answer. What is the status of this "we"? From whence does it come? How is it that a "society" or a "community" with its myriad different individual members or different ethnic groups, each with its own conflicting sets of goals, motivations, and desires, can ever cohere? What *makes* a community?

If I am a stranger or an immigrant in a society, I am perhaps more likely to ask these questions, to ask exactly what it *is* that constitutes the society around me. I have certainly found this in my personal experience as an immigrant in Israel and then England, and it is well known that immigrants often make the best sociologists. The immigrant can't help but stand back somewhat in order to discover a pattern to the "alien" mores and customs in which he finds himself.

Then, too, we often ask whether both selves and societies are real and independent realities in their own right, or whether both are simply convenient fictions for describing the behavior of inherently separate parts. If both are real, which is more primary, the individual self or the society of which it is a part?

These questions are at the heart of both psychology and philosophy on the one hand and of sociology and political theory on the other. They touch each of us as we go about our daily lives. Traditionally they have been represented as the two opposing sides in the perpetual struggle between individualism and collectivism or, in the case of the self, between fragmentation and unity. How we answer them has a deep influence both on our image of ourselves as human beings and on the kinds of organizations we create for binding human beings together in society. I believe that the problem of how to define the self and the problem of how to define society or community cannot be separated. I believe that each individual self is a mini "society" in its own right, and that our insights about the nature of the self have a direct impact on our theories about the nature of society.

In our discussion of how the brain works, we saw that no mechanistic account can explain the unity of consciousness or the emergence of the self. Mechanism stresses the separation of parts and the ultimate reduction of apparent wholes to the functioning of these parts. In the brain these parts are thought to be the individual neurons and their connections. Philosophers who take seriously a mechanistic model of brain function are forced to choose between denying, or simply ignoring, the unity of consciousness or else, like Descartes, arguing that this unity has no physical basis. They are forced to choose between a harsh, "scientific" reductionism or a wholly unscientific dualism.

In his early model of the self, Freud was faced with much the same choice. He had to choose between earlier Christian models of the immortal soul as an unblemished and perfect unity and the more modern, mechanistic emphasis on conflict and division. He chose the latter. The Freudian self can be reduced to three parts, the Id, the Ego, and the Superego. The Id itself can be further reduced to the push and pull of unconscious, instinctual drives associated with sex and aggression.

The Freudian self is a battlefield riven with conflict and dissent. One psychologist has portrayed the Ego as "a referee between a sex crazed monkey (the Id) and a disapproving maiden aunt (the Superego)."[2] Its internal forces are "quantifiable, without reference to any vital integrating agency."[3] That is, the forces that drive and structure the psyche can be reduced to the physical and chemical actions of so many brain parts without taking into account the integrating role of any overall personality. There is no "I," no higher unity, exercising wisdom and control over the disparate forces of this divided self.

Socially, the unity of consciousness and the unity of the self have their counterparts in the unity of society, in the "we" of community and relationship. Modern social philosophers influenced by mechanism have responded to this "we" with the same reductionist instinct that inspires modern philosophers of mind or modern psychologists. They have sought to reduce society to the sum of its parts. Where the separate neuron is the primary unit of mind, or the instinctual force the primary unit of psyche, for the social philosopher the isolated individual becomes the primary unit of society.

On this view, "society," the collective, dissolves into the sum of its individual members. "We," or the community, is just a phantom or a myth for what *appears* to exist when these individuals cooperate, or when their separate characteristics are added together. There is no room in this model for the reality of shared experience or shared values, no room for "we" as a "further fact" over and beyond a haphazard collection of separate "I's."

All mechanistic reduction, at whatever level of mind, psyche, or society that it applies, has its early roots in the atomism of the ancient Greek philosopher Democritus. It is strengthened by the later atomism of Newtonian physics. Democritus' own succinct description of how the atoms behave summarizes why this kind of thinking could never account for the emergence of any genuine collective "I" or "we." "The atoms," he said,

> struggle and are carried about in a void because of their dissimilarities . . . and as they are carried about they collide and are bound together in a binding which makes them touch and be contiguous with one another but does not produce any other single nature whatever from them; for it is silly to think that two things could ever become one.[4]

Atomism in its various forms has exercised a consistent hold over the Western imagination. The primacy of the individual is part of the Western social and political credo. But both atomism, and individualism in its extreme form, have their severe limitations. Both isolate the individual part. Neither can account for any kind of unity, intimacy, or community. Conflict becomes the central metaphor of relationship—conflict, struggle, and collision. Personally, we are left with the fragmented self. Socially, we have Thomas Hobbes's "war of all against all," or the kind of thinking employed in modern defense analysis—thinking that always begins with a perception of others as threats and an analysis of how we can defend ourselves against them.

In both self and society, psychologists and philosophers who adopt the atomistic or individualist model are left asking how we can bring the parts together in any form of even *apparent* unity. What binds all these separate individuals or separate parts of the self together?

Mechanistic philosophers of the sixteenth and seventeenth centuries faced the same question when trying to account for how the physical world "stays in place." Once they had displaced the Renaissance notion that the love of God and the guardianship of the angels drew the world into a smooth, working whole, they had to account for how the parts of their atomistic natural realm could cooperate. Their answer was the new concept of *force*. In Newtonian physics, the colliding atoms of the natural world are bound together by forces, by the push and pull of attraction and repulsion. In the mechanical heavens, these same forces are seen to hold the stars and planets in place. Force, as the new instrument of causality, appeared to offer the physical world a secular cohesion it had never known.

In an atomistic picture of our personal and social lives, this same concept of force takes the form of the power relations that supposedly bind us together. Through the mechanical paradigm, power grips the social imagination. In our everyday lives, power is exercised through a balance of coercion and seduction. Coercion is the "push" of force, seduction is the "pull." Some individualist models invite us to see all our personal and social relations in these terms.

In his model of the divided psyche, for instance, Freud saw the Id as torn between the push of aggression and the pull of sexual attraction. These forces in turn are brought to heel (sublimated) by the Ego through the *power* of Reason. But Freud's vision is a tragic one. In his later work he saw that the mechanical self is caught in a Catch-22 between repression (coercion) and fragmentation. If the conscious Ego represses the Id, the result is neurosis. If the conflicting forces of the Id are allowed free expression, civilization will be destroyed.

Thomas Hobbes and many other social atomists have had the same tragic vision. "The dispositions of men are such," Hobbes said, "that except they be restrained through some coercive power, every man will dread and distrust each other."[5] Earlier still, the Italian political philosopher Niccolò Machiavelli advised his ruling prince to employ a balance of coercion and seduction ("carrots and sticks") to keep his subjects under control. Like Freud with the psyche, both Hobbes and Machiavelli believed that in the absence of these power relations "society" would be torn apart by the conflicting passions of its individual members.

Even in today's modern democratic societies, we can see the push and pull of power relations at play on every level of our lives. We are seduced by advertising and the media and coerced, however subtly, by the economic, political, and legal systems. Years ago Eric Berne's brilliant *Games People Play* exposed the extent of power play that structures everyday personal relations. Yet few of us would accept that in either our personal or our social lives force, or power, leads to any genuine intimacy or sense of community. We cannot force or manipulate people into real relationship. As a model of social cohesion, power is always reductionist. It always sees the individual as primary and separate (like an object) and the social whole as a construct.

Power relations are not the only Western model for why individuals manage to cooperate in society. Some people have suggested that a balance of envy (making certain no one gets too much more than anyone else) keeps conflicting individuals in place. Some suggest that habit and tradition are the glue of society. Others believe that altruism (a concern for others) or some complex calculus of rational self-interest (each of us will see that the interests of the group are in our own individual self-interest) is the cement of society.

But as sociologist Jon Elster says, these are all "partial answers."[6] None on its own—nor even all added together—really gives an adequate account for how or why individuals are bonded together in society. Habit can't account for the creativity of groups, pure altruism can't account for the fact that there clearly is some balancing of individual interests, and the pursuit of rational self-interest never guarantees that we will serve the good of the whole. The famous "prisoners' dilemma" case, in which two prisoners each betray the other out of pursuit of limiting self-interest when a spirit of mutual loyalty would better have benefited both, illustrates this.[7]

Any atomistic or individualist account of the "I" of self or the "we" of community must be a partial account. "I" am the *shared* experience of the $10^{11}$ (hundred billion) neurons in my brain, the $10^{13}$ (ten thousand billion) cells in my body, the $10^5$ (one hundred thousand) genes in my gene pool. I am the shared repository of my biology, my history, my experiences, my culture, and my relationships. "We," the community, are our *shared* experiences, our shared meanings, shared values, and shared projects. We share a communal "story." But in atomism or individualism nothing is shared. The atomistic self or the individualist society is just a pastiche of individual parts or private meanings. As Democritus said,

"It is silly to think that two things could ever become one."

In some important sense, all individualist accounts of society overlook what Freud would have called the Superego dimension, or what Jung would have called the "higher Self." This is the dimension where we find higher values and shared values, where we find what we might call the "transcendent" dimension—a commitment to social and individual principles, which interact in community.

The emphasis in today's very individualistic postsixties society is on qualities represented by the Id—the importance of impulse, of immediate gratification, of antirational, felt response. In the Id society we each do our own thing and we do it because it makes us feel good. Feeling good is often our highest motivation. Feeling good *now.* We rush from partner to partner, from one peak experience to another, from one instant preoccupation (a fad, a TV program, a short article or single-idea book, a drug, a "relationship") to another. We have neither patience nor commitment for things that take time, for what Freud himself referred to as "the slowness with which profound changes bring themselves about in the mind."[8]

In Freud's model, of course, the forces of the Id are pitted against those of the Ego. The Ego acts like a nanny or a policeman, trying to keep the impulsive and chaotic whims of the Id in check. In individualistic social models, the Ego is represented by the various aspects of the power structure—by rationality, rational planning and control, and rational organization (bureaucracy). In the Ego society we are self-centered but rational. We balance our desires with respect to those of others through rational self-interest. We calculate, and we relate through a personal calculus. We manipulate to get our way.

Both the Id and the Ego are opposed to representations from the Superego. In Freud's own reductionist model this is as expected, because the Freudian Superego itself is only the rules imposed by the parents. As Philip Rieff would express it, the Freudian Superego has no "autonomous authority." It has no dimension that in society is represented by the communal "we," by our "collective patterns of thinking, feeling and action," to use Durkheim's phrase.

In what I would call the *true* Superego dimension of society, or in society's "higher Self," we find those collective patterns, Rieff's "right and proper demands superior to competing immediacies, not reducible to nor identical with power."[9] We find our shared experiences (our traditions) and our shared values. It is these that mediate the conflicting demands of reason (Ego) and impulse (Id), that balance between "the rationalizers of technological reason and the orgiasts of revolutionary sensuality" (Rieff).[10] It is in these that we find the true "cement of society" which Jon Elster seeks in vain. In this Superego or higher Self we find a sense of the sacred, more faith than reason, a sense of something beyond the individual, some "further fact," which yet informs and guides the individual. It is a shared way of life, a common set of values, and a loyalty to these.

In any social interaction, neither impulse nor reason is sufficient. In a marriage, for instance, it is not enough either to follow rational self-interest or to act on the demands of impulse or instant gratification. A successful marriage calls on commitment to higher values, on loyalty to some "further fact" which is the institution of marriage itself. A community, too, is held together neither by a complex calculus of self-interest nor by any set of bureaucratic rules. It draws, rather, on its common culture and on the autonomous authority of that culture—on its "collective patterns of thinking, feeling and action" and on their common expression (or common appreciation of that expression). But no individualist or atomistic accounts of society, with their emphasis on Id and Ego qualities, can assign any reality or force to these.

Because atomistic individualism cannot account for any deep unity, intimacy, or communal, shared reality, there has always been a temptation to go to the opposite extreme. Even in early Greek thinking, while Democritus was stressing the divisibility of reality and the isolation of its parts, his predecessor Parmenides had argued in sharp contrast that the All is One. "Nor is it divisible," he said, "wherefore it is wholly continuous. . . . It is complete on every side like the mass of a rounded sphere."[11] For Parmenides, the whole is real, the parts illusory. All multiplicity, difference, and "individuality" are illusions.

All modern collectivist thinkers are in the tradition of Parmenides. They stress unity, the whole, the group, but nearly always at the high cost of minimizing or denying the importance of multiplicity, the part or the individual.

Collectivist models of society or the state stress the oneness of a "higher reality" that transcends, or subsumes, all individual differences. To use Freud's terminology, we might say that they emphasize the Superego to almost the total exclusion of both the Id and the Ego. This collective reality—"inalienable" and "indivisible" in the words of French philosopher Jean-Jacques Rousseau—exists in its own right, apart from the motivations and preferences of its members, and calls upon us to serve it with ultimate obedience. Rousseau called the collective "the general will," saying that it requires "the total alienation of each associate, and all his rights, to the whole community." In Rousseau's ideal constitution, the Social Contract,

> Each of us places in common his person and all his power under the supreme direction of the general will; and as one body we all receive each member as an indivisible part of the whole.[12]

Rousseau's notion of the general will is part of the same tradition that later included Hegel's idea of the state as an "invisible and higher reality"[13] and Marx's inexorable laws of history. All support the belief in social wholes, or social forces, that somehow possess a character and a will of their own over and above the characters and wills of the individual members who make them up. These collective wills are themselves seen as actors, sometimes the prime actors, on history's stage.

Some modern collectivist thinkers have even, like Parmenides, gone so far as to suggest that collective realities are the only true realities, that the whole notion of the individual is suspect. As the English idealist philosopher Bernard Bosanquet expresses it, "All sound theory and good practice are founded on the faith that the common self or moral person of society is more real than the apparent individual."[14]

The clear danger of all forms of collectivism, whether in theories of the self or in theories of society, is that individual diversity (and hence, individual creativity) will suffer in the cause of collective unity or "identity." "I" will be submerged in "we." This, as physicist Leo Esaki points out, is the high cost that Japanese society pays for its almost exclusive stress on the "collective mode."[15]

In collectivist theories of self (those that stress the unity and indivisibility of self), those aspects of the personality that give rise to conflict are subject to repression, or they get fragmented off from the whole. Unwanted impulses or desires are seen as "unhealthy" or even as "alien." We must get rid of them. "If thy right eye offend thee," said Jesus, "pluck it out." This dictum is sometimes carried out literally in modern psychiatric care, where lobotomy is used as a treatment for cutting away unwanted aggressive impulses.

In society, too, those individuals or groups that are not part of the collective can be seen as chaotic and dangerous. They must be repressed, "reeducated," or brainwashed (this happens most frequently in Communist or some Socialist movements and in religious cults), rooted out, cast aside, or even, in the most extreme cases, exterminated. In the new group collectivism of today's radical movements (gay, feminist, ethnic, or even fundamentalist), the errant individual must often be sacrificed for the "greater good" of the collective identity. This has happened with often tragic consequence in the recent practice of "outing," where homosexuals who have wished to keep their sexual preferences secret have been publicly exposed by their fellows. In collectivist theories of education, more concerned with molding children to meet the supposed needs of society than with drawing out the children's own potential, the eccentric, the gifted, or the least able all suffer.

True community cannot be founded by cobbling together a collection of inherently private selves. That is the individual model. But neither can community be founded on the denial or the exclusion of the private or the individual. Neither individualism nor collectivism, in their extreme forms, can give us the kind of public personhood that we described at the beginning of this chapter. Neither can give us a true community of individuals.

In fact neither the extreme individualist nor the extreme collectivist model of relationship mirrors the reality of our daily experience. In our everyday lives we experience a more creative dialogue between ourselves as private individuals and ourselves as public persons, or as members of the community.

Left to ourselves, we do naturally "bunch together." We do experience real moments of intimacy. We form couples, groups, clubs, and societies. We cheer for sports teams and identify with fashion trends. My young children tell me they are very happy at school because each has been elected to a chosen "crowd." Businessmen and industrialists who on the one hand fight for the radical freedom of the laissez-faire market nonetheless find themselves clubbing into lobbies and cartels. Workers pursue their individual rights through trades unions.

At the same time, when the group patterns or group requirements of any collective of which we are a part become too restrictive or too invasive (of our private space, of our personal rights or individual expression), we spontaneously "break out" or "buck the system." We demand a night out, time to ourselves, a corner of our own. We keep a special place in our hearts for the loner, the maverick, or the eccentric. Even within the inner community of the self we cherish a wild streak or a secret corner. At some times in our lives each of us needs to feel that we have done it "my way."

In psychological terms, the roots of both collectivism and individualism might be seen to lie in the original drama with the mother. Collectivism, the All is One, is reminiscent of the child's early fusion with the mother. There are no boundaries, no differences. The child sees itself reflected in the person of the mother. Individualism, with its separation and strife, recalls the adolescent's split from the mother, the heartfelt declaration that each of us is a person in our own right, different and individual. Many of us remain trapped in the first two stages of this drama all our lives. But the mature personality requires some form of adult connecting, where the grown-up child sees himself or herself as fully independent and yet as a member of his or her wider family.

If we are to gain a sense of the part within the whole, of the individual within community, a sense of the part or the individual as made richer, made more itself, *in and through its participation* in the whole, we need a social model that is the equivalent of the mature individual's relationship to his or her family. We need to step outside the constant bouncing back and forth between collectivism and individualism and move on to a new paradigm of relationship that transcends the two. We need to make sense both of the individual *qua* individual and of the individual who transcends himself or herself through a relationship to others and to higher values. We need a modern way to understand the ancient wisdom of Rabbi Hillel's words: "If I am not for myself, who will be? But if I am only for myself, what am I?"[16] I think that the radically new kind of relationship that binds quantum reality into a creative whole can help us to reach this modern understanding.

I want to return to the image with which I began the book, the image of a free-form dance company—each member a soloist in his or her own right but moving creatively in harmony with others. Each soloist stands out as an individual, and yet the dance they make together emerges as a new reality. The dance, the "work of art" or the "production," has an identity of its own over and above the separate and free identities of the individual dancers; each dancer, while remaining distinctly him or herself, acquires a new, a further identity—member of the company.

This dance cannot be described in the old individualist or collectivist terms. It cannot be modeled on any kind of relationship found in mechanistic physics. The dancers are not isolated atoms (particles) bouncing about in the void, nor is their dance an indivisible whole without individual parts (waves). The dance is, I want to argue, a peculiarly "quantum" creation, and can only be modeled in terms of quantum principles of relationship. We have already discussed these principles—wave/particle dualism, indeterminacy, and the holistic, nonlocal correlation of indeterminate parts—in an earlier chapter. Here, I want to "flesh them out," to make them accessible to our social imagination through the image of the dance.

Let us imagine that two such free-form dancers really are a quantum system, like the quantum twins or the correlated photons that we discussed in Chapter 2. Because they are quantum dancers they are capable of getting into a nonlocal correlation when they jump. That is, some of their movements (let's say their arm movements) become synchronized without any force or signal passing between them—they don't touch each other, they never look at each other, and no choreographer is shouting commands to them. This correlation can be positive (both always raise the same arm when they jump), negative (each always raises an opposite arm), or zero (they act independently).

While such correlated quantum dancers might at first seem farfetched, theater director Peter Brook, one of the first to work with the informal dynamics that can arise when groups of actors move or sing together spontaneously, without being tightly choreographed or scripted in singles or pairs, has described something uncannily very similar as an outcome of his "improvisations."

"One thing strikes me as I watch," he says. "At certain moments, two Don Josés, for example, make the same gesture at the same time, on the same phrase, without conniving or even seeing each other. Two identical gestures, as if the music lured them out."[17] Or, again,

> A fantastic exercise: a singer turns his/her back on another and tries to recreate the gesture accompanying the other's singing, without having seen it. You hear the sound, you see nothing, and yet something really can be communicated. Sometimes it really works. Magical.[18]

Returning to our imagined quantum dancers, a physicist would say that the correlation of their arm movements is an "emergent" property of their dance. That is, it only comes into existence as they dance together, when they jump. The correlation of the arm movements cannot be reduced to the personal characteristics of the individual dancers. Nor can it be predicted from their individual characteristics or styles of dance plus any forces acting between them. Neither dancer has any predilection to raise one or another arm, and no forces act between them. The correlated arm movements are a *new* property, one which exists only through the relation of the dancers as they dance (Fig. 4.1).

Through this emergent (nonlocal*) relationship of the arm movements we can understand how it is that the dance itself (or

---

*I.e., in the absence of any local forces.

**FIG. 4.1**

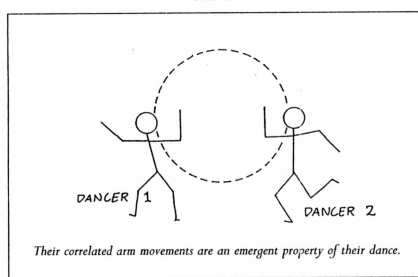

DANCER 1

DANCER 2

*Their correlated arm movements are an emergent property of their dance.*

a ''self'' or a ''society'') can have an identity of its own, how it is a real thing in its own right with its own characteristics and its own ''personality'' superimposed on the original characteristics and personalities of its constituent members (Fig. 4.2). This is not to say, as Hegel might, that the dance or the society possesses some sort of ''super-consciousness'' or ''group mind.'' Rather, the dance is a *repository* of movements and characteristics not possessed by any one member of the dance company. It is a shared style—a shared rhythm or timing in which all the dancers participate as the ''background conditions'' of their individual movements. Similarly, a society (or a culture) is a repository of skills, knowledge, and potential—even of will—not possessed by any one of its members. Like the dance, it has a shared style (shared purpose, shared values). Each of us can draw on this. Each of us can find ourselves a vehicle for the underlying repository's expression.

This idea of a shared repository of skills, knowledge, and potential seems to lie at the heart of David Osborne and Ted Gaebler's call for ''participatory organization'' in *Reinventing Government*. They refer to organizations that draw upon the knowledge and skills of ''frontline workers,'' that acknowledge the ex-

## FIG. 4.2 SOCIETY IS A *FURTHER* REALITY IN ITS OWN RIGHT. SOCIETY, THE "DANCE," IS THE SHARED EMERGENT REALITY IN WHICH THE INDIVIDUALS PARTICIPATE.

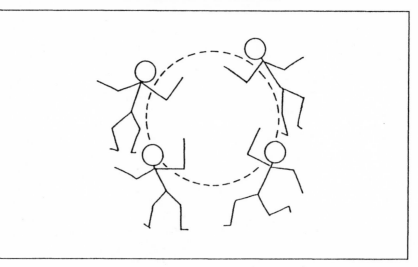

istence of a shared pool of expertise and innovative skill distributed throughout the organization rather than just concentrated in a management structure at the top.[19]

The familiar example of a jazz jam session also might help to make the emergent holism of group identity more accessible. In a jam session, each of the musicians plays as an individual. Each "does his own thing," in his own way. Yet as they play together a pattern emerges that gives some larger, group structure to the sounds produced by each individual. No one musician knows where the piece is going as they play. It "gets composed" or built up *as they play together.*

The French sociologist Emile Durkheim had some similar notion of emergent social identity when he spoke of collective social realities. He believed that "society" could not be reduced to the simple sum of its individual members and their individual psychologies. "A whole," he said, meaning a social whole,

. . . is not identical with the sum of its parts. It is something new, and all its properties differ from those displayed by the parts of which it is composed. Association is not a phenomenon unproductive in itself, consisting merely in bringing into external relation established facts and formed properties. Is it not, on the contrary, the cause of all innovations which have occurred successively in the course of the general evolution of things?[20]

Durkheim saw these innovations occurring through the "coercive" powers of the collective reality. I see them more as a result of an internal, mutually defining relationship between the individual and the group, between the dancer and the dance.

Each member of an emergent relationship finds himself or herself *enriched* by participation in the collective, able to draw on skills or knowledge beyond his or her own individual capacities. Anyone who has ever had a creative idea will have experienced something of its having come from "beyond" the self. The simple act of participating in a brainstorming session has some of this effect. Everyone in the session is "sparked off" by the thinking of the others. The many artists who gathered in Paris during the 1920s or the musicians who wrote in the 1960s seemed to benefit from such collective inspiration, as did the five or six leading figures who collaborated in the founding of quantum mechanics itself.

Personally, I have certainly felt that both *The Quantum Self* and this book have come from beyond the capacities of my own individual consciousness. Most directly I have drawn on my husband's involvement and the ideas that have emerged between us, ideas that neither of us knew previously that we had. I always work better when his attention is on the book as well. Sometimes we even get so close during the writing of these books that we dream each other's dreams. Beyond this, both of us feel that we draw still further on some "spirit" that is abroad in the times (Hegel would have called it the *Zeitgeist*), the thinking and feeling of others only some of which is fully expressed in books or articles or conversation. We often find that we have written something only to discover a short time later that our friends are talking about it or that there is an exchange about it in the newspapers. As I mentioned in the Foreword, in the course of writing this book I have discovered a whole collection of thinkers in different fields trying to articulate the same vision. I think that we all draw strength and inspiration from such participation "beyond the self." We *value* it; when we don't have it, we seek it.

In the emergent relationship between the dancers and the dance, then, we can understand how each of the dancers (segments of the self, or individual members of society) acquires some *additional* identity through the dance. To do so is not entirely without sacrifice. Each sacrifices some of his or her indeterminate *potential*. Each accepts a defining commitment to the style of the dance and thus forgoes certain possible movements that might have clashed. But through this sacrifice of potential each gains something real. Similarly, a man or woman who marries sacrifices his or her potential to marry some other, but through this sacrifice he or she gains a *further* identity as husband or wife.

The dancers *need the dance* to become fully *themselves*. The creative individual needs his society to come fully into possession of his own potential. None of us is truly ourselves in isolation. At the same time, the dance needs its dancers. *The community needs its creative individuals*. Individuals are the "conduits" through which the emergent properties of the community get expressed.

Before they jump, each dancer has a clear set of individual characteristics—fixed (determinate) color of eyes and hair, facial features, physical stature, personal style, etc. None of these is lost in the dance. But each dancer also has an unfixed (indeterminate) *possibility* to raise one or another arm. As they jump, the indeterminate, possible characteristics get into nonlocal correlation, thus giving rise to the new emergent reality, the dance. The determinate characteristics (original identity) remain as they were, sovereign in their own right (Fig. 4.2).

The relation between the dancers' determinate (individual) characteristics and those indeterminate ones evoked through the dance is like the relation between our own private selves and our public personhood. The private and the public are not separate domains. Nor do they simply complement each other. They bring each other into being. Through our involvement in the creatively emergent community, the private and the public aspects of ourselves *define* each other.

In a real free-form dance, the range of possible, unstructured (indeterminate) movements is not limited to arm movements. The range is unlimited, and the emergent dance is a free-flowing, ever-changing, creative movement. The same is true of the jazz jam session. Left to themselves, left to let their nature unfold as it will, I believe that the quantum self or the quantum society for which this dance or the jam session are metaphors could equally well become a free-flowing, ever-changing, creative movement.

In quantum systems, the creative combination of individual identities preserved and enhanced within a collective identity is made possible through the wave/particle duality of quantum reality. The particle aspect is the individual aspect. Its characteristics are determinate. We can locate particles in space and time, we can focus on them and measure them. They are separate and relate mechanically.* The wave aspect is indeterminate. It is a spread of possibilities located everywhere in space and time. We cannot pin waves down; they have no boundaries. Each wave overlaps and combines with all other waves—exists in nonlocal correlation with them. The wave aspect is the creative aspect because it has yet to be fixed ''in place.''

With our quantum dancers, we would say that with their particle aspects they remain separate individuals, while with their wave aspects their identities overlap and combine to form the additional, ''higher'' reality of the dance. In the self, we would say that with their particle aspects the individual subselves remain separate and maintain their own identities (their own personalities), while with their wave aspects the subselves combine to form the higher unity of the ''self.'' In society the same duality would give us individuals fixed in their separate and meaningful identities, *and* it would give us ''community'' with its further collective identity. Unlike the traditional individualist or collectivist models, neither the individual nor the collective entity is more primary or more real. In the quantum realm, both waves (relationships) and particles (individuality) are equally real and equally primary. They coexist.

Harvard philosopher John Rawls, in his now classic *A Theory*

---

*Strictly speaking, ''particles'' can be of any size—subatomic, the size of tables, or larger. Their defining properties are their determinacy and their mechanical relationships, not their size.

*of Justice,*[21] gives a very powerful description of the wave, or group, aspect of our place in society with his notion that "justice is fairness." Rawls uses a very concrete fable to illustrate the impersonal point of view we must adopt when deciding what is fair for all in society. In his fable, each of us is cloaked in what he calls a "veil of ignorance." That is, we don't know what place we occupy as individuals in society—we might be among the best off, among the worst off, or somewhere in between. From this position of ignorance we must decide how much *all* individuals should receive in the distribution of rights and goods. This fable is very like the well-known children's party technique of allowing one child to slice the birthday cake but allowing the other children first choice of which piece each shall take. Presumably, in both Rawls's fable and the birthday cake example, we will opt for the fairest possible distribution in hope that whatever we end up with won't be too bad.

The trouble with Rawls's emphasis on the impersonal distribution principle is that it doesn't offer any counterbalancing discussion of the good often achieved for the whole of society by some people pursuing a more personal or self-centered point of view. We all sometimes gain from the wealth created by certain swashbuckling entrepreneurs who seek nothing more lofty than their own gain. Similarly, the whole process of biological evolution itself gains a certain dynamic from various species' trying to get the best situation for their own.

At least one major political philosopher has suggested that a more double-edged understanding of individuals and their relationships is necessary to transcend the old conflict between "the claims of the group and those of the individual." In his latest work, American philosopher Thomas Nagel proposes that all individuals contain within themselves two standpoints, the personal and the impersonal. From the personal standpoint, the individual looks out for himself, for his own interests. But from the impersonal standpoint he empathizes with, even identifies himself with, the needs and desires of others.

"Without the impersonal standpoint," Nagel says, "there would be no morality, only the clash, compromise, and occasional convergence of individual perspectives. It is because a human being does not occupy only his own point of view that each of us is susceptible to the claims of others through private and public morality.''[22] I would suggest that Nagel's personal standpoint follows from the particle aspect of our quantum nature, while his impersonal aspect is made possible by the wave aspect.

With our particle aspect we stand apart and experience life from our own point of view; with our wave aspect we are literally taken up by, woven into, the being of others and of all that surrounds us. It is this duality that allows us to ''contain and be contained by others without our personal truth ever being wholly isolated or exhausted.'' It is this duality that makes us persons and members of a community. It accounts for the sense of fulfillment, the sense of truly coming home to ourselves, that each of us feels when we genuinely become part of something larger than ourselves.

## Society's Wave and Particle Aspects

The wave/particle dualism in self or society is a powerful metaphor on which we can build a new model to transcend the individualist/collectivist divide, but I believe it is more than a metaphor. I believe that human consciousness really is quantum mechanical in its origins, and that the mechanics of this quantum consciousness *literally* give our minds, our selves, and our social relations both a wave aspect and a particle aspect. The wave aspect is associated with our unstructured potential, with our spreading out across the boundaries of space, time, choice, and identity. The particle aspect gives us our structured reality, our boundaries, our clearly defined selves, our ordered thoughts, our social roles and conventions, our rules and patterns.

In each of the cases just discussed, in the dance, in the self, or in society, the creative collective that emerges through the relationship of the wave aspect, or the free, indeterminate characteristics of the individual parts, is what physicists call an emergent ''relational whole.''[23] Such wholes cannot be reduced to the sum of their parts. They are built upon the indeterminate, ''possible'' characteristics that those parts *come to have* only when they relate. In these entities, relationship *evokes* reality. It literally ''calls out'' a possibility latent within the situation and actualizes it. In the physics laboratory we see this relational holism in the emergence of nonlocal, correlated polarizations when two "separate" photons are measured. In the social realm, we can see it very clearly at work, for instance, in the dynamics of crowds, or mobs.

We know from countless examples that crowd or mob behavior is often peculiarly threatening. Individuals caught up in mobs will feel things and do things that on their own would never occur to them. The mob environment may elicit feelings and behavior that participating individuals never knew they possessed and may afterward try to disclaim. "I wasn't really myself when I did that," they will say or, "I just got carried along." I believe that this is literally true and that it has clear legal implications.

A mob is a social "relational whole" of the most visible sort. The characteristics and behavior of the mob emerge from the pool of latent (indeterminate) characteristics of its individual members, but until these latent characteristics are correlated they have only a phantom reality. It is the coming together of the individuals, the eerie correlation of hitherto unrealized violent tendencies (possibilities) that gives rise to the mob and its passions. The responsibility borne by the individual who participates in a mob is just that—he is responsible for joining a mob, for allowing himself to become part of something that will "bear him along."

One articulate young football hooligan* I spoke to said that being "borne along" was his real reason for attending football matches. "You go and you get caught up in the crowd," he said. "You become part of the crowd and it gives you something. It feels good. Maybe it's as close as most of us will ever come to finding God." Each of us can recognize this same motivation to get in touch with something larger than ourselves in our own less violent social liaisons. Emile Durkheim argued that the feelings evoked by such crowd experiences were in fact the basis of our belief in and notions about a sacred reality. Speaking of Australian Aborigines' participation in an ecstatic tribal dance, he said,

*A soccer fan who gets involved in vandalism and riots during or after the game. Such "hooligans" are, unfortunately, a familiar feature of European sport.

It is not difficult to imagine that, having reached this state of exaltation, man no longer knows himself, and feels himself dominated, carried away by a kind of outside power. . . . And since at the same time all his companions are transfigured in the same manner and express their feelings by their cries, their gestures, their attitudes, all proceeds as if they were transported into a special world . . . the world of sacred things.[24]

Mobs are, as I say, social relational wholes. Durkheim called them "social facts." So, too, I believe, are families, groups, and nations. Each carries within itself a spiritual or sacred dimension possessed of a power often for good but sometimes for evil. We gain access to this power through our social rituals—our shared values, shared attitudes, shared activities. But relational holism need not be restricted to social entities. We might expect to find it as a dimension of reality wherever quantum systems combine. Where these quantum systems are conscious, we would find it giving rise to art (the drawing together of perspectives or colors into an emergent whole), music (the emergent pattern arising from the separate notes), or conceptual entities (language, truth, goodness, beauty). All are subjective emergent wholes greater than the sum of their parts. Through the consciousness of the individual who perceives them they become emergent realities.

Relational holism can be seen as the key creative (or, as in the case of mobs, destructive) leap that gives rise to new, further social realities at every stage of the evolution of human relationship. This evolution begins in the quantum system of the brain itself at the level of the neuron cell-wall molecules or the neural ion channels. It begins with the indeterminate jiggling of these molecules, which when correlated (through electrochemical stimulation) gives rise to the quantum unity* that is the physical basis of consciousness.

This living quantum whole, with its various "excitations" arising from (but not reducible to) sensation, experience, memory, instinct, emotion, and the genetic code, would be the first con-

---

*The Bose-Einstein condensate.

scious collective or relational whole, the first "I." That "I" might contain within itself several subselves, several smaller regions of quantum identity. We could picture the self as a system of emerging whirlpools, each discrete and individual at its center but overlapped and merged with others at the periphery. The great American psychologist William James developed in the last century such an idea of the self as a "seething cauldron . . . where everything is fizzling and bobbing about in a state of bewildering activity."[25] Yet in all this "fizzling and bobbing about" there is always a center, the emergent reality of the "I" that conjoins the indeterminate characteristics of the subselves.

Beyond the always shifting boundaries of the "I," relational holism draws the unfixed aspects of the self into ever wider circles of creative relationship—the intimate partner, the family, the group, the nation, "humanity" itself. Other emergent qualities draw the self into creative relationship with nature, art, culture, values, spirituality, and, ultimately, "god consciousness," or the cosmos. At every evolutionary stage the healthy self has the potential to become part of, and to define, a larger, collective entity—to derive new layers of its self-definition from that further reality—while remaining always itself, a dancer in the dance.

We will have reason to discuss relational wholes, or "relational holism," throughout the remainder of this book. They are perhaps the most basic conceptual building blocks of a quantum society. They bear on almost every aspect of social reality, altering fundamentally our concept of relationship and the nature of community. In the next few chapters, I want to explore the social conditions in which true community can flourish, the radical change in perspective necessary to foster these conditions, and the creative potential of doing so.

# 5. Freedom and Ambiguity: The Foundations of Creative Community

*The more [rigidly] connected are the elements of a system, the less influence they will have on the system as a whole. . . . The more [rigid] the connections, the more each element of the system will exhibit a greater degree of "alienation" from the whole.*

Von Foerster's Theorem*

I recently did a filmed interview for British television. The producer making the film had very fixed ideas. She had conceived a particular scene that she wanted to film, and my participation was meant to justify its inclusion. She assigned me a given role in her script and wanted me to speak to this chosen part. She told me what to wear, where to sit, and even what to say. I felt manipulated and couldn't "get into" the part. As the camera rolled, I found that my thoughts escaped me and my contribution was lifeless. Had the producer understood Von Foerster's Theorem and its relevance to human situations, she might have seen the inevitability of this outcome.

There are several other daily situations in which

---

*"Von Foerster's Theorem on Connectedness and Organization: Semantic Applications," Benny Shanon and Henri Atlan, *New Ideas in Psychology*, Vol. 8, No. 1, pp. 81–90, 1990.

we can see this same dynamic at work. There is the example of a couple having a row, for instance. When both members of the couple are angry, they remain fixed in a position of heated confrontation. If one member begins to feel slightly better, he or she almost inevitably zeroes in on how the other should change, too. There will be advice to "cheer up" or "smile." But such cues have the opposite effect. The alienated partner experiences them as a demand and becomes even more fixed in an entrenched position. It is usually far better to leave the mood of the other untouched and do something indirect.

Again, when there is tension in a group because of the behavior of one or more members, there is a temptation for others to focus on that behavior and to demand that it change. But when a disturbed or upset person is focused upon, he or she becomes fixed in that behavior pattern, becomes locked into a role, and remains alienated from the group. Things usually work out better if the group can learn to communicate indirectly, can learn just to *be* together, without focusing on any one member. Then a sense of their group cohesion usually reemerges.

We can see in all these cases—the interview, the intimate relationship, and the group—that overdetermination, or coercion, isolates, or "alienates," the individuals involved from the group as a whole. All are familiar to us, and each is an example of Von Foerster's Theorem at work in social situations. The theorem tells us that there is an inevitable tension between the focused upon, the fixed, the particular, or the individual and the sense of the whole or wider relationship.

This tension reminds me of the Uncertainty Principle that we find operating at the level of quantum reality. In the original Heisenberg formulation of the Uncertainty Principle, we are forced to choose between measuring a quantum entity's position (its particle aspect) or measuring its momentum (its wave aspect). The particle aspect is the individual, the particular, located in space and time. The wave aspect is the spread out (nonlocal), the holistic. We can never know both at once. In Von Foerster's Theorem, we are forced to choose between tight structure and a sense of the whole.

There is a vital, internal relationship between Von Foerster's

Theorem and the Uncertainty Principle as it applies either to quantum systems or to human social situations. Together, I believe they have something important to tell us about the conditions necessary to foster a true sense of community. Both bear directly on our ideas about freedom and the relation of free individuals to groups or wider social realities of which they are a part. Indeed, I believe that linking the two gives Von Foerster's Theorem an important quantum twist that makes it more applicable to human social situations.

Von Foerster's Theorem itself was originally formulated to describe the behavior of cybernetic systems, systems that achieve a kind of internal, homeostatic control through the free exchange of information between the parts (e.g., our central heating systems, most information technology systems, and many of our own biological functions). To our mechanistically conditioned imaginations, the theorem seems to go against common sense. It argues that if one element of a system is rigidly determined, rigidly fixed in place, that element loses its relationship to the whole. It becomes isolated, or "alienated." Conversely, the theorem says, the less fixed or the more uncertain the behavior of any element of a system, the greater will be its influence on the system as a whole. Our mechanistic intuitions run counter to this.

Our impulse is that the wholeness of a system is best determined by "nailing" its pieces in place. Too often we see the elements of situations as being related like the cogs and wheels in a machine. Each cog, we think, plays its most efficient role in the machine if it is securely fitted in place. The television producer thought she could best ensure the integrity of her film by rigidly determining each participant's role. She felt that if she allowed the individual contributions to be more spontaneous they might be "woolly," might not add up to a coherent whole. In our social relations we often feel that rigid social structure, firm political control, detailed bureaucratic rules, or a strong emphasis on well-defined social roles is necessary to ensure a smooth working of the whole. Both Von Foerster's Theorem and the Uncertainty Principle argue that such thinking is at the least one-sided.

Our impulse to nail things in place in order to ensure the workings of the whole is an impulse to control. This impulse is related to the way in which power grips the mechanistic imagination. Things, and individuals, are perceived to be related through force, and force is an instrument of control.

It is true that we *can* force people into a kind of tight structure, just as chemical forces can push molecules into a tight crystal lattice. A highly ordered society built on rigid social roles and a well-disciplined army with its tight hierarchical structure are both social equivalents of a crystal lattice. A crystal does have a high degree of order, yet each molecule in a crystal lattice is inherently separate. It is bound to the whole only through external forces. It is, in Von Foerster's terms, "alienated." Individuals in the equivalent, tightly controlled social structures have very little influence on the groups of which they are a part. Such controlled group cohesion is bought at the expense of individual freedom, expression, or influence, and the individuals concerned often rebel. Both sedition and corruption are more common in dictatorships than in democracies. The economies of dictatorships are usually less healthy because tightly controlled individuals show less initiative and work less hard. Later, I will argue the same is true of rigid bureaucracies. In psychotherapy, the same dynamic exists. If a therapist imposes a tight theoretical interpretation on the patient's remarks, the patient finds himself less able to speak freely. He or she becomes alienated and cooperates less well.

Liberalism, of course, arose precisely to react against such tight and uncreative overcontrol. Foremost among modern political philosophies, it tries to take on board the insight that imposed structure and rigid control constrain and alienate the individuals who are being controlled. It reacts strongly against any restrictive social rules, whether these are laid down through tradition or by a bureaucracy. Liberalism's central metaphor is the Social Contract, a fictional device through which free and independent individuals agree to a very minimal amount of necessary cooperation. Beyond this minimal cooperation, each individual is free to cooperate further or to compete in his or her own best interests.

Liberalism rests on a mechanistic model of society in which the individuals (atoms) are primary and their relationships are secondary and external. "I" precedes "we." In an ideal liberal society,* each of us is seen as a separate self, like the separate elements in a cybernetic system. Each of us is given maximum freedom to pursue our own ends, and guaranteed the right of free communication with all the other separate selves. More than that, no one of us is meant to think of anything but our own (selfish) purposes. None has any "higher" motive to serve the good of the whole. Indeed the whole, society, exists solely to facilitate our individual ends and purposes.

Yet liberalism asserts that somehow, by some mysterious homeostatic (cybernetic) process, we free, communicating individuals will achieve a "natural harmony of interests." Each, through pursuing our "private vices," will unwittingly, and almost mysteriously, serve the public benefit. Such thinking was the basis of Adam Smith's belief that an "invisible hand" guided the selfish pursuits of a free-market entrepreneur. Almost despite himself, Smith's ideal entrepreneur serves interests beyond his own.

> He generally, indeed, neither intends to promote the public interest, nor knows how much he is promoting it. . . . He intends only his own gain, and he is in this, as in many other cases, led by an invisible hand to promote an end that was no part of his intention.[1]

Indeed, for Smith and other classical liberals, it was *crucial* to liberal theory that the individual pursue his own interests. This was essential to their definition of liberty. As John Stuart Mill said, "The only liberty that deserves the name is that of pursuing our own good in our own way":

> . . . unless men are left to live as they wish 'in the path which merely concerns themselves', civilization cannot advance; the truth will not, for lack of a free market in ideas, come to light; there will be no scope for spontaneity, originality, genius, for

---

*In the pages that follow, I am not speaking specifically of modern political liberalism, with its egalitarian concerns, but rather of the classic liberalism articulated in the seventeenth and eighteenth centuries. All the many forms that liberalism has taken since preserve the basic features I am discussing here, though the specific political ends sought have varied considerably.

mental energy, for moral courage. Society will be crushed by the weight of 'collective mediocrity'.[2]

Adam Smith believed that positive harm might result if his free-market entrepreneur set out to serve the public good rather than his own. "By pursuing his own interest," Smith argued, "[the entrepreneur] frequently promotes that of society more effectually than when he really intends to promote it. I have never known much good done by those who affected to trade for the public good."[3]

For classical liberalism, individual liberty is the highest value, the pursuit of individual ends the ultimate purpose. "Society," to the extent that it is considered at all, is a neutral arena in which the drama of this pursuit is played out. As sociologist Robert Bellah and his colleagues say about the American brand of liberal individualism, "For many of us, 'freedom' still has the meaning of a right to be left alone."[4] The highest "social good" is seen as a delicate balance of mutual self-interests, as the preservation of a right to be left alone.

Liberal individualism embraces a very particular view of the self. Crucial to this model is what John Rawls calls "the distinction between persons," or what Robert Nozick terms "the fact of our separate existences." This is the essential atomism of the liberal credo. And not only is each self separate from all other selves, but the liberal self is also prior to all its ends and associations. "I" am not a mother or a wife. "I" am not my relations with my family and friends; "I" am not my social roles, my professional commitments, or my national loyalties. "I" am not even my achievements or my moral qualities, nor my character. All these things are "contingent" characteristics, not essential qualities of my self. "I" am a separate and individual person and all my rights follow from the mere fact that I am.[5]

Like atoms that can link together in molecules but always remain unchanged in themselves, liberal individuals can participate in society, they can even influence society, but society cannot influence them. As Bellah and his colleagues comment on the American social crisis, "The individualist assumptions of our culture lead us to believe that we can live as we choose, using the big institutions—the agencies of the state, the companies or organizations that we work for, the schools we attend—for our own ends, without being fundamentally influenced by them."[6] Such assumptions lead us directly to Christopher Lasch's "culture of narcissism."[7] We care for nothing but ourselves, our own goals, and our own emotions. Things or people other than ourselves are just a means to our feeling good. We become absorbed in ourselves, lost in ourselves, only to find that, ultimately, we are condemned to ourselves.

Individualist "society" is not a community in the sense that we discussed in the last chapter. It is not an emergent whole in which the parts take on new qualities through creative participation in something larger than themselves. Adam Smith's "invisible hand" is emergent only in the loose sense that unexpected or unforeseen general patterns or laws suddenly appear or become noticeable. For instance, timekeeping is an emergent property arising when the parts of a clock are assembled in the right way. The apparent law of "survival of the fittest" in Darwinian biology or the laws of thermodynamics that describe the behavior of mechanistic systems are also emergent in this sense. But the individuals (or the atoms) that participate in these emerging patterns don't themselves change. The individual, *qua* individual, does not evolve. The cog in a clock remains a cog. A dog goes about the business of living his individual life irrespective of whether or not he happens to be among the fittest. The individual remains as he was and he remains alienated from (ignorant of and unmotivated by) any emerging pattern—a mere games player or actor in some script of which he is unaware. Similarly, the liberal individual is to no extent defined through his relation to others or to the workings of the invisible hand. In any such mechanistic model, "society" can always be reduced to the sum of us individual members and our private interests.

As liberal selves we do not discover or create ourselves through community. Our spontaneity, originality, and genius have no meaningful context in which to unfold and flourish. We are dancers without any dance. On the contrary, as the British philosopher Isaiah Berlin says, "It follows [from this kind of thinking] that a frontier must be drawn between an area of private life and that of public authority.''[8] He means by this that liberal thinking imposes a rigid boundary between the two. Most individualist thinking extends this frontier to exclude the *whole* of ''public life.'' As rugged liberal individuals we stand alone, often pitted against society. We are not true members of any community. The community is no *part* of us. It doesn't get *inside* us to become part of our identity as individuals. We are alienated from the whole (and, I would argue, from important aspects of ourselves).

The liberal individualist self, prior to all its ends and associations, can never be other than it is. Our freedom to pursue our own ends and to be ourselves can never raise us to anything more than ourselves. It is an *external* freedom. In this sense, I would argue that as liberal individuals we have a limited freedom. We have what Isaiah Berlin calls a ''negative freedom,'' or a ''freedom from,'' but not a ''freedom to become.'' We have actuality but no potentiality.

The founders of liberalism had a great vision. They were not themselves selfish or self-seeking men. They were not indifferent to personal growth or to the social good. But they saw that growth and that good as best served by free, atomistic individuals serving themselves. They were reacting against the very traditional societies of their day, in the same way that Newton had reacted against the constraints of Aristotelian and medieval science. Just as Newton personally had his deep Christian faith to sustain him, thinkers like Locke and Mill had their own deep, inherited social values. They were, as Rabbi Jonathan Sacks has pointed out, ''able to take for granted a high degree of shared morality and shared belief, without having to reflect too much on the institutions that sustained it.''[9] But those traditional, shared values were no more built into their political philosophy than the Christian faith was built into classical physics.

Both classical physics and classic liberalism in themselves were to leave their later followers in a spiritual vacuum. Classical physics has no spiritual dimension; liberal society, per se, has no ''higher Self,'' no Superego dimension. It has a commitment to individual freedom but no commitment to a shared social reality. Under its influence there is the danger that we become hostile to all shared tradition. Its critics have also argued that liberalism's freedoms are often of more use to its winners than to its losers.

There is, of course, an opposing tradition of political thinkers who reject the notion that the individual is best served by serving himself. Known in its modern form as "communitarianism," this tradition cherishes a very different myth. Communitarians argue that human beings have always lived in groups and that we need to do so. We have, they believe, no essential or fixed nature in ourselves but we develop our attitudes and our characters through being members of the group. "We" is more fundamental than "I." This tradition's major historical thinkers are Hegel and Marx, perhaps even Aristotle, and include today such postmodern philosophers as Wittgenstein.

For communitarians, one important political task of society is the education, or reeducation, of its poorly adapted members until they become able to find their "true" fulfillment, or true "freedom," through participating fully in the communal life around them. As Hegel says,

Everything that a man is he owes to the state; only in it can he find his essence. All value that a man has, all spiritual reality, he has only through the state. . . . The state is not there for the sake of the citizens; one could say, it is the goal and they are the instruments.[10]

Though Hegel meant the "state" to be the politically organized ethical life of the community rather than any bureaucratic power, it is easy to see how his communitarian bias might be used to justify coercion. Other communitarian thinkers have substituted "church" or "Reason" or "social blueprint" for the "state," but all agree there is some best truth or most fulfilling way to live toward which we all should aspire if we are to realize ourselves. To a great many of us, this kind of thinking raises very obvious danger signals. As Isaiah Berlin says, this "freedom" to do what others think best is "no better than a specious disguise for brutal tyranny":

. . . to manipulate men, to propel them towards goals which you—the social reformer—see, but they may not, is to deny their human essence, to treat them as objects without wills of their own, and therefore to degrade them.[11]

Neither the external "freedom from" of the atomistic liberal individual nor the communitarian's spurious freedom to become some specified something or other that others think "good" for me leads to the desired goal of a free individual who, through his freedom, achieves both a creative sense of the whole and a creative sense of himself—a creative sense of belonging. The liberal individual belongs to nothing and no one. He is an isolated island unto himself. The citizen of the communitarian, or collectivist, society has nothing but a sense of the whole. His alleged "freedom for" is really a sacrifice of all personal freedom.

## Complementarity and Quantum Community

True community cannot flourish in either a liberal individualist or a communitarian society. Neither, I would argue, can meaningful, spiritually rich, freedom. Both individualism and collectivism are mechanistic social models. Individualism sees particles, or atoms, as primary and reduces all social wholes to a sum of those parts. Collectivism is more a wave model of society. For the collectivist, the part is always subsumed in the whole. The pattern, or the wave, is primary. Neither a particlelike nor a wavelike model of society can give us a collection of free individuals who lose their alienation (and enrich themselves) through a greater participation in the whole.

For any creative sense of community, we need some flexible duality of particlelike (individual) and wavelike (relational) characteristics, some combination of what Thomas Nagel calls our personal and impersonal points of view. No mechanistic social model can give us this duality, or combination. I believe that a quantum model can, but to understand exactly how we must return to the Uncertainty Principle (i.e., the quantum version of Von Foerster's Theorem) and quantum indeterminacy and see how these might be extended to a social application.

The Uncertainty Principle, we remember, tells us that if a quantum system is wholly determinate in one aspect it is wholly indeterminate in the complementary aspect. We can know its position *or* its momentum, its energy *or* its time, etc., but fixing one in place unfixes the other. It is also possible for some aspects of the quantum system to be *partly* determinate and *partly* indeterminate. As an analogy, for instance, if we face in any one direction, we have a very clear vision ahead (determinate aspect), no vision at all behind (indeterminate aspect), and partial, blurred vision off to the sides (aspects that are partly determinate and partly indeterminate). In physics itself, for instance, an electron in an atomic orbit has a definite energy but both its position and its momentum are in a partially indefinite state. These are spread out in a fuzzy, doughnutlike ring around the nucleus.

The complementary pair that is of interest for our social model is the particle/wave pair. The particle aspect is the individual or the particular aspect, located in space and time. Each separate system has a fixed number of quanta, not shared with others. Particles have fixed characteristics and external, one-to-one relationships. The wave aspect is the indeterminate, the relational, spread out all over space and time. Waves overlap and combine, they have holistic interference effects, they get inside each other. They get into nonlocal correlations and have internal relationships. In the two-slit experiment (Chapter 2), we found that if we measure the particlelike aspect, we destroy the wavelike aspect, and vice versa. Similarly, if a member of a football team is concerned only with his own performance (his particlelike aspect), he will have little concern for the performance of the team as a whole (his wavelike aspect). In language, we see the same distinction between the tight either/or constraints of bureaucratic or legal prose and the evocative both/and use of metaphor or poetry, where several meanings are conveyed simultaneously.

In different contexts or under different circumstances, quantum "things" can oscillate between the particlelike and the wavelike qualities or balance midway between the two extremes. In human terms, as we shall see from the many examples that follow in a moment, we would say they are a balance of Nagel's personal and impersonal points of view. They have both. But when we disturb these dualities, when we focus on them and measure them, we can only see one aspect or the other. This is the nub of the Uncertainty Principle. When we measure a quantum entity, what we do is extract one of its potentials.

If we were to draw a parallel between the effect of measurement on a quantum system and possible relationships between the individual and society, we would find the measured state akin to the familiar pull between individualism and collectivism. If we focus on one quantum entity, we force it into an extreme "individualist" state, a particle state alienated from the nexus of internal relationships. If we focus on the group characteristics, we force the entity into an extreme "collectivist" state, a state where relationship is all and individuality is lost. Neither extreme measured state preserves the particle/wave balance (the individual/collective or personal/impersonal balance) that distinguishes quantum reality.

In personal communication we see the same thing. If we cross-examine someone in a legalistic fashion, we do the equivalent of measuring him—we pin him down to a particlelike either/or statement. If, by contrast, we just listen to someone sympathetically, allowing all his associations to flow without interference, we leave him in a more wavelike state. In both these extremes we lose something valuable—free association and nuance in the first, and definite information in the second. A free-floating brainstorming session seems to capture the middle ground between the two extremes. When we brainstorm, we throw out all the relevant ideas that occur to us, but we also focus on some of them as better than others. There is a definite theme to the session, though no tight constraint.

It is when both the particle and wave aspects are *somewhat* indeterminate that we find the potential for an emergent quantum "community." It is here that we find *both* the capacity for free "individual expression" and the capacity for creative "belonging," coexisting in the same entity. The individual in the brainstorming session makes his contribution as an individual, but he is often inspired to do so by the overall intellectual energy of the group.

The creativity of a quantum system, then, like that of the free-form dancers in the last chapter, rests on its duality. It rests on the fact that neither all the individual characteristics (particlelike potential) nor all the relational characteristics (wavelike potential) are fixed in an extreme position. The freedom possessed by such a system is an *internal* freedom. The whole system, its identity, its "character," its expression, and its receptiveness, is somewhat unfixed, inside and out. It is in a state of flux and becoming, like the free-form dance we have used as our central image or the music generated in a jazz jam session. This internal freedom *thrusts* the system into "community" (into the dance or the music), where the indeterminate, relational aspect acquires its identity as it relates.

For the self, internal freedom is a freedom to become something more than oneself through a relationship with others or through experience. Where the liberal self can never become more than it is, where it is never influenced by its associations, its relationships, its institutions, or its career (because it has only *external* freedom), the quantum self is in a state of constant becoming. This internal freedom to become underlies the quantum self's motivation to get into relationship or community or to have experience.

It is through such liaisons or experiences that we grow, that we real-ize our own further potentialities. As Martin Buber says about the defining qualities of an I-*thou* relationship, "Relation is mutual. My *thou* affects me as I affect it. We are moulded by our pupils and built up by our works."[12] This potential for being affected is why the overintellectual man often seeks out a woman of feeling, or a disorganized woman seeks a practical man. This is why people who find it difficult to communicate often become writers or why people with unresolved personal problems often become counselors or therapists or priests. We feel a passion for those liaisons or careers or experiences that will allow us to grow in the direction that our own unexpressed potential makes us feel we need to grow. Our internal freedom (our indeterminacy) makes this possible.

In our everyday social lives, so much of what we value and do depends upon a vast array of indeterminate, or ambiguous, social cues and relations. A smile, a nod of the head, a joke, an offer of help, or an invitation to some activity may all have a whole range of potential meanings and potentially different outcomes. When a woman chooses to wear a particular dress for an occasion, she isn't necessarily intending some calculated effect, like seduction. She may just have a "feel" for the occasion or some inarticulate feeling about herself. If my son invites a friend round to play, he seldom has any fixed agenda in mind, nor is any required. Parents who try to overorganize their children's play often simply block creativity.

Freudian psychology, being mechanistic in its whole model of the person, always looks for *the* meaning hidden in ambiguous gestures or remarks. The analyst acts like a detective, trying to uncover or unmask these hidden meanings. But many of our meanings genuinely are ambiguous; we create them as we go along. If this were not so, there would be no scope for creative response or the emergence of creative social relations.

We saw this with the quantum dancers in the last chapter. It was their unfixed gestures and characteristics that allowed them to get "in sync" (in correlation) with the other dancers, their indeterminate characteristics that allowed them to discover themselves through the dance. The same is true in the real life of the photons in the laboratory experiment—their "introduction" to each other only results in a correlation of some of their characteristics because those characteristics were indeterminate, or ambiguous, in the first place (Chapter 2).

Remarkably, most of the founders of quantum physics—Dirac, Bohr, Pauli, de Broglie, Schrödinger, etc.—thought that elementary particles exhibit some kind of agency or "choice" in their behavior.[13] As Nobel physicist Freeman Dyson expresses it, "The processes of human consciousness differ only in degree but not in kind from the processes of quantum states which we call 'chance' when they are made by electrons."[14] But David Bohm has added the insight that such agency can affect the environment only to the degree that the quantum entities are in a wavelike state, that is, to the extent that they are still in an as yet indeterminate state (indefinite space-time position, but definite energy and momentum).

It is the definiteness of the momentum that gives the system its definite causal properties, i.e., its ability to act. A wave has a definite momentum, like a wave washing up on the beach and crushing the sand beneath it. Similarly, a team that is acting together, or a crowd of people who have got into synchrony, are both wavelike states (as is Japanese society), and all these are more effective than a group of uncoordinated individuals (who have definite position, but indefinite momentum). In physics itself, coherent (wavelike) laser light has far more intensity than the incoherent light given off by ordinary lamps. As we saw in Chapter 3, the kind of quantum system most likely to exhibit the *most* agencylike properties is a Bose-Einstein condensate[15]—exactly the type of quantum system I have proposed is the most likely physical basis for consciousness.

The upshot of this is that consciousness owes its own unified agencylike properties (i.e. its free will and responsibility) to its having an indeterminate and wavelike physical substrate. The implications for a quantum society are that its members will exhibit the highest degree of creative effectiveness (the *least* alienation) if they are not too constrained by rules and procedures. "Pinning down" the members of an organization with bureaucratic rules deprives them of any agency or power over the outcome of their actions—hence the "alienation." A looser, more flexible organization may allow a "group agency" or sense of purpose to emerge. This, from a quantum angle, is the kind of conclusion we reach when we interpret Von Foerster's Theorem in the light of the Uncertainty Principle.

Theater director Peter Brook has described a dynamic in his own work that seems to illustrate all this exactly. Speaking of some people's wish to see theatrical sets designed in advance by great painters or sculptors, Brook notes that such fixed designs, imposed from the outside, very often constrain, even take the life out of, a production. "What is necessary," he says, "is an incomplete design; a design that has clarity without rigidity; one that can be called 'open' as against 'shut.' . . . A true theater designer will think of his designs as being all the time in motion, in action, in relation to what the actor brings to the scene as it unfolds. The later he makes his decisions, the better."[16]

In the same way, our many ambiguous communications and ambiguous feelings are like a background pool of potential relationship that can in time be made more precise according to circumstances. The myths, legends, literature of a culture serve a similar purpose. They are shared but many-faceted, full of potential meaning and multiple interpretation. They can be reinterpreted as people grow and change. Ambiguous or double meanings in words and sentences allow shared meanings to form. Such ambiguity and duality is basic to letting creative social relations emerge. Each one of us who is both an individual and a lover, an individual and a parent, an individual and member of some chosen group has experienced this double-edged, individual/relational quality of human identity. I have argued that it is made possible by the quantum nature of our consciousness.

## Fostering Community

The question posed by any appreciation of the role ambiguity and indeterminacy plays in our lives, the role of our inner freedom to become, is how can we foster it? What, concretely, can each of us do in his or her own life to realize the full potential of our two-sided (individual [particle]/relational [wave]) nature? How can we live our lives so that those lives are both fully our own and at the same time fully realized through a relationship with others?

At the simple, surface level, whatever our inner nature, we have seen that we must allow ourselves a high degree of freedom. The institutions we evolve in society must not impose too much external control. It is impossible to *impose* a sense of the whole on individuals. We can't prescribe creativity, we can't define hard and fast rules for achieving emergent holism. We can't, as the collectivist does, hope to achieve a sense of community through some grand social blueprint that legislates a rigid economic doctrine or a "socially conducive" educational policy. But avoiding these pitfalls only assures freedom in its external sense. It gives us, at best, the freedom cherished by liberal individualism, a freedom that does not in itself nurture community. External freedom is a necessary but not a sufficient condition for fostering community.

To foster community, we must also nurture our internal freedom, our freedom to become through relationship with others.

This internal freedom is linked to our indeterminate characteristics, to those aspects of ourselves that remain unfixed and hence free to get into a correlated relationship with others. In quantum mechanics, we have seen that a balance of determinacy and indeterminacy (a particle/wave or individual/relational balance) is threatened by measurement. If we want to preserve the creative ambiguities of social situations, and thus nurture those dualities that allow us to participate in emergent social reality, we must avoid to some extent the human equivalent of measurement.

For the physicist, measurement is a way of *looking* at the situation. We have seen that in the quantum realm, the way we look at a system partly determines what we will see. If we design an experiment to measure the wavelike properties of a quantum entity, a wave is what we will see. If we design an experiment to look at particlelike properties, we will see a particle. I believe the same is true of our dualistic personal properties. I believe that the way we *look* at ourselves partly determines what we then are, what we become. Our perceptions give rise to our reality. As the sociologist Peter Berger says about identity, our identity is a *constructed* reality. "Men not only define themselves, but they actualize these definitions in real experience—*they live them.*"[17]

And the way we look at ourselves, the way we perceive ourselves, is defined by our attitudes. *Attitude,* I believe, is the human equivalent of measurement. *The attitudes that we adopt in any situation partially determine how that situation will unfold.* Our attitudes hold the key to the amount of internal freedom we can enjoy. They hold the key to the kinds, and the extent, of relationships that we can enjoy, and hence to the kinds of communities that we build.

If, for instance, I adopt the attitude of liberal individualism, I see myself as separate from others. I look for those qualities about myself that set me apart, I concentrate on those goals I want to pursue for my own benefit. I avoid making binding commitments. I face situations and relationships with the attitude, "What's in this for *me?*" In quantum terms, I see myself as a particle. I become "fixed" in my identity, removed from the nexus of personal and social potential that was mine before I adopted this attitude. As a separate individual (a particle), I will not be an effective member of any community. This is a danger for politicians, actors, and other public figures. They are widely seen in a particular role, and they get stereotyped. If they accept this rigid definition of themselves, which is what originally made them famous, they lose the potential for further kinds of relationship and growth.

If, on the other hand, I adopt a collectivist attitude, I lose sight of myself as an individual altogether. I am given over to those qualities acquired through my relationship with others, I have my goals decided for me, I seek only the benefit of the group, with which I am wholly identified. I face situations and relationships with the attitude, "Please tell me who I am. Please give me the feeling that I exist." I sacrifice myself to my relationships. I am overtaken by my social roles. In quantum terms, I see myself as a wave. I become "fixed" in a position of anonymity, removed from the nexus of personal and social identity that were my potential before I adopted this totally altruistic attitude. As an overwhelmed or wholly passive nonperson (a wave), I have nothing creative to contribute to the community.

In more concrete terms still, imagine that I am about to go to a meeting. If I go into the meeting with a fixed agenda in mind, a proposal that I want the group to approve, I am adopting an attitude that I want to use the meeting for my own ends. I will be less open to the suggestions of others, more "alienated" from any group dynamic that evolves during the meeting. I cut myself off from the group's potential. If, on the other hand, I go to the meeting with no ideas to put forward, if I adopt the attitude that I will allow myself to be carried along by the will of the group, I will make no creative contribution as an individual. Neither attitude is conducive to the emergence of a community spirit in the group.

A sense of "community," whether an emergent relationship between members of an intimate couple or a larger-scale relationship on a group level, can only emerge where the individual members stand poised to allow the indeterminacy (inner freedom) of the situation to unfold. They must, like the physicist who aids the unfolding of physical reality in his laboratory, stand ready to become midwives to the unfolding social reality of the situation. This poise is incompatible with the striking of any fixed, extreme attitude. It is inseparable from trust—trust in the unfolding potential of the situation and trust in oneself as an individual to "ride" with that situation in a skilled way.

There are many daily social situations in which the kind of poise I am trying to describe plays its creative role. Any mother who has ever held a crying baby will know that she neither tries to force the baby to stop crying (adopting an attitude of control) nor allows herself to become hysterical (adopting an attitude of surrender). The mother stands poised and alert, ready to cope, trusting in both her own skill and a kind of inarticulate bond between herself and the baby from which a solution to the crisis will emerge. Any friend or lover, faced with another in great distress, stands poised in the same way, listening creatively, *being there* creatively, while neither controlling nor surrendering to the other's distress. A good psychotherapist uses this skill every day. It is, as we shall see later, the essential basis of genuine dialogue and the emergent realities to which dialogue can give rise.

In less intimate social situations, this same poise strikes a creative balance between more fixed attitudes of control at the one extreme and total receptiveness at the other. It draws out the potential latent within any meeting between two or more people. Any successful negotiation, commercial or political, depends upon it.

If a negotiator is too fixed in his attitude toward the negotiations, the process of trying to find a mutually satisfying outcome usually breaks down. If, on the other hand, a negotiator is too receptive, he is likely to be overwhelmed by the wishes of others. Even in the wider sphere of something like a national election, a successful democratic outcome—i.e., a true consensus expressing the wishes of the community—depends upon there being a significant number of voters who go into the election campaign with open minds. If voters are too rigidly fixed in their political attitudes, the election becomes a shouting match. If they are too receptive, they may not bother to vote at all, or they will simply go along with whatever the opinion polls tell them most other people are planning to do.

I believe that the potential latent within all social situations arises from the fact that we are quantum persons, from the fact that each one of us has indeterminacy (inner freedom) built into the structure of our consciousness. Whenever we meet, or wherever we join together in groups, communities, or nations, this indeterminacy gives us the potential to get into correlation with one another, the potential for the emergent holism of the group to express itself.

Like the quantum dancers we met in the last chapter, the community is our dance. The community is that further social reality through which we both express ourselves and creatively discover more of ourselves. The extent to which we participate in and contribute to the dance is a matter of attitude. There is a whole range of conflicting, fixed attitudes that in one way or another pulls us out from our creative bond with the community, and an equally wide range of personal qualities that nurture a poised stance toward community.

I think, for example, of the crucial distinctions among obsession, love, and promiscuity. An obsession fixes us in place, it pins us (and the other) down and ties us as slaves to its object, whether that object is a game of chess, a job, or a person. Indeed it always treats the other as "object." It is "particlelike." It pulls us out of wider relationship and is seldom creative. At the opposite extreme, promiscuity leaves us wholly unfixed, undefined, all adrift in a sea of possibilities to no one of which we meaningfully commit ourselves. It is "wavelike." Only love is poised and alert, accepting of the reality of that with which it is in relation, committed to that relationship and ready to let it evolve, ready, if necessary, even to let it go rather than to pin it down and turn it into an object.

The same is true of the triad: fanaticism, loyalty, and following; or of involvement versus either standing separate or letting oneself be overwhelmed. Both loyalty and involvement are poised attitudes that show a commitment to emerging relationship rather than adopting either a fixed (and hence object-directed) stance or no stance at all. I have thought of many more such attitude triads (Table 5.1), and I am certain that readers could think of others. Doing so, and reflecting upon them, can be a form of meditation and even perhaps a vehicle for transformation.

We can see now, through reflecting on the chart of attitudes

*Ideal* (handwritten)

## TABLE. 5.1

*Cons. servative* (handwritten, left margin)  *Liberal* (handwritten, right margin)

| RIGID ATTITUDES LEADING TO ALIENATION FROM GROUP | POISED STANCE COMPATIBLE WITH CREATIVE FREEDOM | RIGID ATTITUDES LEADING TO ALIENATION FROM SELF |
|---|---|---|
| OBSESSION | LOVE | PROMISCUITY |
| FANATICISM | LOYALTY | FOLLOWING |
| ROLE PLAYING | CHARACTER | ANOMIE |
| HABIT | STYLE | IMITATION (TRENDINESS) |
| IDEOLOGICAL | OPEN-MINDED | TOTALLY RECEPTIVE |
| PAROCHIALISM | COMMITMENT/DEDICATION | VACILLATION (PERMANENT INDECISION) |
| SINGLE-MINDED | SENSE OF VALUE | RELATIVISM |
| OBSERVATION | PARTICIPATION | SUBMERGENCE (BEING TAKEN OVER) |
| SEPARATE | INVOLVED | OVERWHELMED |
| INDEPENDENT (MY WAY) | DIALOGUE (OUR WAY) | CONFORMITY (YOUR WAY) |

INDIVIDUAL (PARTICLE) ◄─────── / ───────► GROUP (WAVE)

*moderate   see Aristotle —* (handwritten)

leading to alienation, what each one of us can do as individuals to help transform ourselves and thus the society in which we live. We can see what each one of us can do to live out the inner truth of Von Foerster's Theorem or the Uncertainty Principle. The transformation from a mechanistic, atomistic society of selfish individuals where each pursues his or her own way in isolation to a quantum society with a vibrant sense of emergent, creative community requires a transformation of personal attitude. It requires that each one of us, to some extent, let go of our fixed perceptions, our habits, our obsessions, our rigid ideologies, our single-minded pursuit of personal gain, and our parochial devotion to our own corner. It requires, instead, that we stand poised and alert, poised to let our inner freedom (our indeterminacy) give rise to the unfolding, common reality of self and community.

At the same time, Table 5.1 shows us that the poise necessary to foster creative involvement is compatible with—indeed requires—commitment and personal style. It requires character, loyalty, and a sense of value. The letting go of *everything* that we believe thrusts us into the wavelike extremity. As Ulysses sailed past the sirens, he did not close his ears, but he did ask his sailors to tie him to the mast. Some habits are necessary, some attitudes are good and necessary to have. They are like the straps that bind us to the mast so we are free to listen.

There is no firm set of formal rules that can tell us when to hold on and when to let go. We learn through experience and reflection and through the experience of others. It is a form of tacit knowing, something that itself emerges through our inner freedom and our creative dialogue with others. It is to that dialogue that I want to turn next, to the poised attitude we must cultivate toward the ways and truths of others, to the indeterminacy and multiple possibility that lie at the heart of all our relations with others and that ultimately define both our selves and our place in the wider community.

# 6. The Many Faces of Truth

*No finitely describable system, or finite language, can prove all truths.*

    *Truth cannot fully be caught in a finite net.*

Gödel's Theorem

My two children have had very wide exposure to both Christianity and Judaism. They have often seen the inside of both churches and synagogues and have repeated the prayers familiar to each faith. Very recently we took them to the Far East, where they spent weeks visiting Buddhist temples and hearing the Buddhist account of the universe. I have told them this exposure to many faiths is good for them because, as the Jewish mystical tradition teaches, God has ten faces and they should know as many of these as possible. But the children are unhappy with this. They say it confuses them. As children do, they want things to be more simple, more singular. Often they ask me, "But which is *true?*"

The urge to know the simple, singular truth troubled the mathematician Kurt Gödel as well. He wanted to know whether we could ever write a mathematical statement (an equation) that could contain all truth within itself.* Gödel was a Plato-

---

*More accurately, whether we could ever formulate a rich or interesting mathematical system that could contain proofs of all its theorems.

nist. He believed that a Realm of Forms, or world of pure truth, does exist, but through his work he proved that no human mathematician ever could sum this up, or express it. Gödel's Theorem[1] tells us that whatever we try to say in human language, in *any* kind of language, will always be partial. This is as true of the "languages" of Christianity and Judaism as it is of the language of mathematics. All must content themselves with being partial expressions of a "higher" or a "deeper," and ultimately an inexpressible, truth. There are analogies to this in the new physics.

Physicists, for instance, can calculate only with values measured in finite terms—definite amounts of mass, density, gravitational force, and so on. But the equations of General Relativity that we use to describe the present universe break down—i.e., they yield infinite answers with which we cannot calculate—when we try to describe the *origin* of the universe, the Big Bang itself. So cosmologists can give only an approximate description of the beginning of things. The same thing happens in quantum field theory. If, instead of contenting ourselves with describing the *manifestations* of (or appearances on the "surface" of) underlying reality ("excitations" of the quantum vacuum), we try to describe underlying reality (the vacuum) *itself*, we are destined to fail. Whenever we try to measure how much energy the vacuum itself contains, we get an infinite answer with which we cannot calculate. Our physics always breaks down en route to the Absolute. Thus physicists, too, must content themselves with partial truth.

Gödel's Theorem lies at the heart of a very modern philosophical revolution. It has sweeping implications for the nature of mind and for all the mind's claims to ultimate truth. Because such claims are themselves at the heart of how we live our lives and how we relate to others, the theorem touches the very foundations of our life together in society. It touches the reasons that we give and the reasons that we seek for our actions, and the reasons that we are prepared to accept from others for their actions. It touches the foundation of value and our tolerance for the values of others.

A passion for simple, singular truth is not unique to children. Ever since Moses returned from Sinai with his message of the One God, the West has rejected polytheism in all its forms. Moses' God was a jealous God, who demanded recognition and obedience to the exclusion of all others. The story of this God was for His people the "true" story, His commandments the "true" laws and practices. Those who observed them were chosen, those who did not were outsiders. Both Christianity and Islam adopted this same monotheistic model in their still stronger claim to being bearers of the one, true faith, the one, revealed truth. For the Christian or the Muslim, those not in possession of this faith were worse than outsiders. They were pagans and infidels. *Heretics*.

The Greeks, too, though they recognized a whole pantheon of gods and celebrated diversity through their tragedies and dialogues, had the ideal of a single, perfect truth, accessible to all—with Apollo's help—through the faculty of reason. Plato's Realm of Forms is a paradise of perfect truth, each Form being the one, true model for every imperfect thing of its kind in this world. Knowledge of the Forms is innate, truth a matter of unambiguous recall. The ignorant slave boy in the *Meno* can give the correct answer to Socrates' geometrical problem because his soul has prior knowledge of all geometrical form. He has the same innate access to a knowledge of virtue, and any two (or ten) slave boys would agree, after suitable reflection, upon the nature of true virtue. "All nature is akin," says Socrates, "and the soul has learned everything."[2]

Moses and Plato are the two pillars of our Western preoccupation with singularity and simplicity, of our passion for clarity and our distrust of ambiguity. Their common vision of a single, ultimate, and knowable truth became the spiritual foundation of Christianity and later still the intellectual foundation for the philosophical and scientific revolution of the seventeenth century. It was out of this revolution that the modern age was born.

Modernity began with Cartesian doubt and the rejection of all claims to a revealed, or given, truth. Descartes wanted to rid his mind of every received opinion, every notion formed on the basis of what others, or even his own senses, had told him, and rebuild all his ideas "on foundations that would be wholly mine."[3] But the foundations that Descartes was to claim as his own were very narrow and exclusive. They were to rest on pure Reason alone, pure, disembodied, singular Reason.

Descartes' sharp divide between mind and body drew an equally sharp division between Reason and experience. Reason, like the mind, was unified and indivisible. Reason would never lie or confuse. The truths revealed by Reason were "clear and distinct." Experience, on the other hand, the whole realm of the body and its senses, was by nature divisible and fragmented. Experience was often ambiguous, the senses could mislead.

Newtonian physics was in the same spirit as Cartesian philosophy. It, too, drew a sharp distinction between mind and body. But for Newton, the focus of concern and the value judgment was exactly the opposite of Descartes'. For Newton, it was the physical realm that was unified—by simple determinist law. Newton's physical world was all of a piece. Everything was in its place and clarity reigned supreme. In mechanistic physics everything is as it seems, three simple laws provide the foundation for all truth, and there is only one reality. The logic of this physics, like the logic of Moses' monotheism or Plato's theory of truth, is a straightforward either/or logic. It is what the poet William Blake described as Newton's "single vision," or "Newton's sleep."

While Newton's physics isolated mind and conscious experience as we know it, at the same time his new mechanical science became a general model for much of the thinking about mind and truth that was to follow. Whether through Cartesian dualism or through mechanism, modernity adopted the value judgment that unity and singularity are "good" and that disunity, or fragmentation, is "bad," or at least "untamed."

In psychology, Freud adopted a mechanical model of the self. He also adopted the modernist split between unity and disunity and the value judgment that one is good and the other bad. All irrationality, all fragmentation or plurality, was consigned to the Freudian Id, the dark and tempestuous region of the psyche. In this scheme it was the task of Ego to shine the pure light of (unitary) Reason on the Id's murky depths. "Where Id was," Freud hoped, "there Ego shall be."

Today, computationalism, or artificial intelligence, follows the same pattern. "Mind" is thought to be like a good computing machine. Mind is rational, rule-bound, deterministic and hence predictable, and is subsumed by a unified, narrowly defined Reason. The fragmented or the plural, the ambiguous, the unpredictable, cannot be accounted for in cognitive science. As I have argued elsewhere, *consciousness* cannot be accounted for.

The upshot of modernity's emphasis on the "goodness" of unity, or singular truth, is that much of our experience and our humanity are left out. The plurality, the contradictions, the ambiguities, the conflict, and the tragedy arising from conflict are either denigrated or denied. And Reason itself, the basis for what we take to be objective, is itself narrowed. The objective becomes the measurable, the repeatable, the determined, something that can be located within an absolute frame of reference, like Newton's space-time coordinates.

Modernity began as a movement to cast aside all divine, revealed truth and all outside authority. Descartes rejected everything that did not present itself clearly to his own mind. And yet modernity in fact simply replaced one God with another. It replaced Moses' One God with a singular, clear and distinct, either/ or Reason. This Reason, and the narrow interpretation of objectivity that goes with it, was just as authoritarian and intolerant, just as eternal and immutable as divine, revealed truth. In the pure light of Reason there could be only one reality, only one way of looking at things, only one truth. And the *laws* of Reason determined what this would be.

It is because we have remained under the spell of monotheism, or the belief in one, simple, singular truth, that the history of the West is a history of intolerance and bloodshed, a history of crusades and holy wars, of inquisitions, of guillotines, pogrom and holocaust. "One belief, more than any other," says philosopher Isaiah Berlin,

> . . . is responsible for the slaughter of individuals on the altars of the great historical ideals. . . . This is the belief that somewhere, in the past or in the future, in divine revelation or in the mind of an individual thinker, in the pronouncements of history or science, or in the simple heart of an uncorrupted good man, there is a final solution. This ancient faith rests on the conviction that all the positive values in which men have believed must, in the end, be compatible, and perhaps even entail one another.[4]

In the twentieth century, during the period known as "high modernism," this faith in one, dominating voice of reason came together with the singularity of mechanism as never before. "The belief 'in linear progress, absolute truths, and rational planning of ideal social orders,' " says Oxford's David Harvey, gave rise to a "celebration of corporate bureaucratic power and rationality, under the guise of a return to surface worship of the efficient machine as a sufficient myth to embody all human aspirations."[5] Hitler's machines for killing were the most chilling embodiment of a surrender to overrational mechanization and to the kind of bureaucratic, instrumental reason that produces it.[6]

But in splitting off the multiple from the unified, by denying the plurality of experience in favor of a vaunted Reason, modernity became a Janus/Bifrons, a two-headed monster. It carried within itself the seeds of its own destruction. In repressing that side of our experience that is rich with nuance and variety, in denying the validity of different ways of looking at things, including the intuitive and the irrational, the worship of unity acted to cripple and distort the capacity for multiplicity. The multiple became the dark and tortured underside of modernity itself, the Id within the modern psyche. As Baudelaire said in his famous definition, "Modernity is the transient, the fleeting, the contingent; it is the *one half* [emphasis mine] of art, the other being the eternal and the immutable."[7]

On the one side of the full modern experience, then, we have extreme, imposed unity. On the opposite side we have the nightmare of fragmentation. Wherever there is such a split, it is inevitable to find the one side rising up in strong reaction against the other. In ancient Greece the split was between Apollo, god of reason, and Dionysus, god of revelry, the passionate, and the irrational. But the Greeks saw both as valuable and held them in balance. In the late eighteenth century, a more extreme, one-sided reaction against reason and absolutism came from the Romantics, who celebrated diversity, nature, and again, the irrational. In our own times, in the late twentieth century, similarly one-sided rebellion has come from the cultural movement dubbed "deconstructive postmodernity."

This movement has all the characteristics of a child rebelling against a too-stern father.*

Deconstructive postmodernity claims to be a rejection of the values held by modernity, but as David Harvey argues, it is really a rejection of the one half of modernity in favor of the other. Deconstructive postmodernity is a rejection of unity and Reason in favor of disunity, fragmentation, the irrational, and the unpredictable. Where modernity (Descartes and Newton and their followers) said there is only one objective point of view, only one truth, only one Reason, the deconstructive postmodern writers and philosophers say there are many points of view, but all are subjective, many "truths," but all are relative. They say that Reason is a myth and that all the constructs of Reason are mere facades. At the center of postmodernity is the fragmented self, fragmented consciousness.

In *Alice's Adventures in Wonderland,* the Dodo introduces Alice to the notion of a Caucus-race. He marks out some space with no particular shape, various creatures stand around in it "here and there," and in quite random fashion some of them run about. "There was no 'one, two, three and away,' " notes Alice, "but they began when they liked, and left off when they liked, so it was not easy to know when the race was over."

The Caucus-race is an irrational way to run a race. It is a race with no center, no rules, no authority, and no purpose. The Dodo decides that as there are no criteria by which he can choose a winner, they may as well all be winners, but that decision, too, is arbitrary. They might just as well all be losers. The very concepts "winners" and "losers" seem a matter of convention dictated by the Dodo's whim.

The Caucus-race seems to me an apt metaphor for the decon-

---

*The term *postmodern* has a confusing range of recent applications. In one sense, it is used to describe a whole range of thinking that in any way sets itself against or beyond modernity. In its more narrow, and more popular sense, it has come to be associated specifically with the kind of nihilistic and relativist thinking of certain continental "deconstructionist" philosophers. I am using it here in that more narrow, popular sense which, properly speaking, must be given the unfortunately unwieldy label, "deconstructionist postmodernism." (See Charles Jencks, 1992.) In later chapters I will use the term *postmodern* on its own to refer to anything that comes after the "modern."

structive postmodern reaction to modernity's passion for single, immutable truth. Deconstructive postmodernity is a style of thinking in which there are no winners or losers but only unrooted conventions, all equally valid (or invalid) and all equally devoid of meaning. I believe it is a style that has undermined sound common sense in nearly every field of contemporary culture.

In architecture, the postmodern is what one architect calls a "junkyard of the past," a senseless pastiche of clashing styles, each borrowed without rhyme or reason from some earlier era and now jumbled together as a parody of all meaning. Sometimes this is done with a sense of fun, more often with outright cynicism, but the result is that it usually confuses and disorients the observer.

In fiction, deconstructive postmodernity offers novels that lack either clear plot or character, nonnovels that parody their own form and content. One example is Umberto Eco's novel *Foucault's Pendulum,* which a fellow novelist has characterized as "babble and gobbledygook." It ends with Casaubon's conclusion that "I have understood. And the certainty that there is nothing to understand should be my peace, my triumph."

Postmodern (deconstructive) science tells us that there are experimental data that have no ultimate meaning—we simply interpret them as we like. Supposedly objective criteria for deciding what is real and what is not "are temporary resting places constructed for utilitarian ends."[8] Deconstructive postmodern theology suggests we can have religious "feeling" though we believe nothing (Don Cupitt's *Taking Leave of God*). And deconstructive postmodern history reduces the historical to a collection of mere nostalgias. "History" breaks down into an infinity of "histories," each from a different perspective, or interpretation. There is no progress, or at least no further progress. In his *End of History,* Francis Fukuyama claims, "There will be neither art nor philosophy, just the perpetual caretaking of the museum of history."[9]

All things postmodern in this deconstructive sense share a feeling of the "used-upness" of past form with a sense that nothing new can ever happen. We exhaust the past only to discover there is no future, and betwixt and between we dash about like the creatures who run the Caucus-race, seeing the contradictions that fragment the rules of play rather than looking for the unifying theme that might give the race some point.

Where modernity's belief in simple, absolute truth and constant, linear progress was one of essential optimism, the philosophy of deconstructive postmodernism is equally one of despair. It is rooted in the deep rejection by many recent professional philosophers of all claim to certainty or grounded belief. It is rooted in their lack of confidence that they have *anything* worthwhile to say, in their loss of faith in the historical task of philosophy. These philosophers have led the way in taking others down the path of relativism and despair.

Philosophy was once the queen of the sciences. Her pursuit was associated with a love of wisdom and her insights enlightened the most profound developments of the human spirit. The results of her study have quickened the imaginations and colored the perceptions of ordinary people as they have gone about their daily business, even those who have never heard of philosophy. When something goes wrong with philosophy, something goes wrong with us all.

Today very many things are wrong with professional philosophy. In the world of English philosophy, we have become very used to philosophers lowering their sights, restricting their debate to the twenty-one different uses of the *word* "good" instead of to any discussion of the Good in itself. This restriction followed from the linguistic analysis of Ludwig Wittgenstein, from his claim that we could never get beyond the "language game" in which we are immersed to say what words mean outside the game. There is *nothing,* he claimed, that we can say outside the game, and only convention within it.

In continental Europe (i.e., Europe excluding the British Isles—an important distinction in European politics and culture) and in America—very much influenced by Continental thinking— things are, if anything, even worse than in British philosophy. According to several influential Continental philosophers— Jacques Derrida, Jacques Lacan, Michel Foucault, Jean François Lyotard—philosophy is dead. They say we are living at "the end of philosophy."

"Philosophers usually think of their discipline as one which

discusses perennial, eternal problems, '' says American philosopher Richard Rorty, ''as an attempt to isolate the True or the
Good. . . . But there is no interesting work to be done in this
area.''[10] There are, Rorty concedes, local instances of truth or
goodness, but nothing general or useful can be said about what
they have in common. There are no such things as Truth or Goodness, and philosophers who search for them are in error.

Philosophy, the deconstructive postmodernists argue, must
end its quest for objective universals because there is no ''outside''
platform on which to stand in looking for them. No Realm of
Forms nor Sinai nor Newtonian space-time coordinates. ''There
is,'' as Jacques Derrida is famous for saying, ''nothing outside the
text.'' All platforms on which we might stand are inside some
system of language or convention. Richard Rorty would replace
Plato's philosopher king, who perceived the eternal Forms, with
the ''informed dilettante,''[11] an all-rounder familiar with various
language games but committed to none.

Rorty's views mirror those of his contemporaries who have
made it their declared aim to undo philosophy, to chip away at
the foundations of thought and of the self who thinks, to shock
and undermine—to deconstruct—the beholder's sense of reality.
Reality itself, any thing or being, independent of human fabrica-
tion, is just what they say is at best inaccessible. In place of truth
or reality, we have instead only limited human discourse, the systems of belief and acts of interpretation that each of us makes from
within the prison of our own language or culture. To challenge
these supposed ''truths,'' to deconstruct the assumptions on which
they rest, is the task of the age.

All the deconstructive postmodern philosophers trace their an-
cestry back to Nietzsche, and they have been greatly influenced by
the German existentialist Martin Heidegger. While Nietzsche is
best known for his declaration that God is dead, his claim that
Reason, too, is just a pawn of myth and illusion had equally deep
philosophical implications. It led to his own assertion of the end
of Man, the end of Descartes' Enlightenment ideal of a rational
human nature. Heidegger defined Man as a ''there-ness'' (Dasein),
a being who is always ''thrown'' among the language and the
conventions of the everyday world.

The upshot of the deconstructive postmodern rebellion is the claim that every value is equal to every other value. Nothing is real or natural or authoritative. Everything is up for interpretation—goodness, God, literary meaning or merit, artistic or architectural standards, even gender (particularly a big issue in feminist circles). And because there is no "center," no single natural or superior "language-game" to which others can be referred, the postmodernist tells us we should have a bit of them all.

Deconstructive postmodernity is always eclectic, a value that has found its way into the syllabus of many schools and universities where a core curriculum has been dropped in favor of "any little bit of what you fancy." Reading the scripts of Marx Brothers films or Ealing comedies is equivalent to reading the novels of Jane Austen. The oral tradition of an African tribe has the same cultural weight as a play by Shakespeare. This insistent eclecticism lies at the heart of much of what passes for the "politically correct" at American universities.

Deconstructive postmodernists rank eclecticism as a supreme value, but by the criteria of their own system, I would argue that it, too, is surely just another value in the potpourri of values in which we dabble.

In David Lodge's novel about the impact of postmodern thinking on academic life, a pithy north-of-England industrialist dismisses a woman instructor's "de-constructionist" ideas with the outburst, "I've never heard such a lot of balls in all my life!"[12] Allan Bloom's best-selling *The Closing of the American Mind* drew much the same conclusion. But the deconstructive postmodern rebellion against any kind of objective standard for truth or value cannot be shrugged off lightly.

"The issues raised by de-constructive postmodern thought are neither frivolous nor trivial," argues one Oxford University lecturer in Continental philosophy. "They are the central philosophical problem of today. We may somehow get beyond the post-modern, but we can never just go back to the old ways of thinking."[13] Many people in recent years have tried to do just that, as is shown by the recent return to fundamentalism.

The rise of Islamic fundamentalism in the Mideast and of an

equally vehement fundamentalism among some Christians and Jews in Europe and America is a natural reaction to too much cultural uprooting and uncertainty. A similar strain of intolerant cultural fundamentalism has greeted the wilder excesses of the politically correct syndrome in America. Human beings must have some minimal anchor for their thoughts and beliefs, some minimal pattern for how they should live their lives. I believe this need is embedded in the dynamics of consciousness, in a physics that leads the conscious mind to form an ordered, coherent whole from the data of experience.

The possible quantum basis of consciousness that I described in Chapter 2 is an example of what the physicist Ilya Prigogine calls a "self-organizing system." Such systems exist throughout the physical world—the simplest, everyday example being the whirlpool that we see in the bathroom basin each morning. The driving force behind the basin whirlpool is gravity pulling the water down into the hole. But in doing so, it draws the water into a self-perpetuating pattern—all whirlpools in the Northern Hemisphere, large or small, rotate in the same direction. And they continue their rotating pattern even though the separate water molecules being sucked into the whirlpool are different at every moment. It is the pattern, not the actual water molecules, that persists.* In the same way, the pattern of each individual's human body persists even though every body totally exchanges every atom of its substance within a period of seven years. It is our physical (self-organizing) pattern that persists, not the actual substance of our bodies.

Self-organizing systems take chaos from the surrounding environment and pull it into a dynamic, ordered pattern. In the case of the body, we take disorganized food and energy and transform it into substance for the body's pattern. In the case of the conscious mind, the chaos is the plethora of information that bombards the brain at every moment. Mind takes the chaotic information and draws it into a pattern—if this is cultural information, into a

---

*Whirlpools are not quantum self-organizing systems. They are simple, "classical" systems, but they exhibit the same self-organizing dynamic as a quantum system like a Bose-Einstein condensate in the brain would.

"worldview," or a lifestyle.[14] When an individual's, or a group's shared, worldview is threatened by too much chaos, the physical pressure to consolidate is enormous. The result, in consciousness, is a temptation to rigid fundamentalism of one sort or another.

Newton's third law of motion stated that "For every action there is an equal and opposite reaction." Culturally, we have become stuck in this mechanistic pattern. An assertion that there is only one truth is met by the reaction that there is no truth, a claim to the Absolute is met by a cry for the relative, and that in turn is met by a renewed cry for the Absolute. This is traditional either/or thinking. It forces us to choose between distorted extremes. If we are to get beyond it, we must find some radically different grounding for truth and value and for our access to them. The Dodo's Caucus-race cannot be won or lost on the old terms.

In his novel *The Gift of Asher Lev,* Chaim Potok writes a dialogue between the artist hero, Asher Lev, and his father, a fundamentalist Jew. Describing the roots of his art in the complexities of human experience, Asher says,

> I call that ambiguity. Riddles, puzzles, double meanings, lost possibilities, the dark side to the light, the light side to darkness, different perspectives on the same things. Nothing in this whole world has only one side to it. Everything is like a kaleidoscope. That's what I'm trying to capture in my art. That's what I mean by ambiguity.

His father replies,

> No one can live in a kaleidoscope, Asher. God is not a kaleidoscope. God is not ambiguous. Our faith in Him is not ambiguous. From ambiguity I would not derive the strength to do all the things I must do. Ambiguity is darkness. Certainty is light. Darkness is the world of the Other Side.[15]

The response of Asher's father is familiar. He is a passionate and honorable follower of simple, singular truth. But Asher's plea for a more ambiguous perception is something fresh. Asher is not a deconstructive postmodernist. He is not a relativist or a follower of no truth. He is himself a deeply religious man whose kaleido-scopic vision springs from the deepest sources of human creativity. In ambiguity he sees possibility, not confusion. In multiplicity he sees a truth that cannot be summed up easily, or perhaps not at all. His vision is that of an artist, but it is a vision that I believe is accessible to us all. Through cultivating it I think we can transcend the old absolutist-relativist split, the extreme choice between one truth and no truth.

## Many-Faceted Quantum Truth

I want to return now to quantum reality, to the possible quantum physical nature of human consciousness and to the kind of truth the human mind can glean from its creative dialogue with the world in which it finds itself. It is in that dialogue that I think we can find the vision that can make artists of us all.

Quantum reality, we must remember, is very different from the world that Newton presents for our imagination. Newton's world is rule-bound and deterministic. There are definite physical laws for the behavior of particles and their interactions. If we know these laws and the initial state of the system we can predict ac-curately where the system is going. There is no uncertainty, no ambiguity. Everything is as it seems. Quantum reality, on the other hand, is indeterminate and probabilistic.

Newton's world is a world of either/or. Light is either a series of waves or it is a stream of particles, a particle is either here or it is there, either now or then. Quantum reality, however, is a world of both/and. The Schrödinger wave equation that describes things in the quantum realm is a plethora of often mutually con-tradictory possibilities, a spread of coexisting contradictions all equally real, all equally true, in their potentiality. Light is both a wave and a particle; a particle can be both here and there, now and then. Quantum systems are spread out all over space and time.

The difference between Newton's world and the quantum world is an important metaphysical difference. It affects our whole attitude to reality. The metaphysics of Newton is a metaphysics of actuality. It is about the here and now, the tangible, and actualities do exist in a world of either/or. But quantum metaphysics,* like earlier Aristotelian metaphysics, is a metaphysics of potential. In both Aristotle's philosophy and in the vision of reality underpinning Heisenberg's Uncertainty Principle, the *potential* of a thing or a system is an important part of its description.

A quantum system's potentials, or "propensities," all coexist simultaneously and all play a part in the evolving dynamics of the system until some act of measurement selects one of them to be the next actuality. This two-level description, actualities and possibilities, is metaphysically much richer than the single-reality world of mechanistic physics.[16] And though quantum possibilities are relative to the environment (i.e. context dependent), they are not subjective or whimsical. Quantum mechanics is an extremely accurate description of them. As we shall see, they are crucial to our model for relationships in a quantum society.

Both the many possibilities carried by the quantum wave function and the duality of quantum reality have strong implications for our need to get beyond the impasse that divides modernity (single truth) and deconstructive postmodernity (no-truth). Each may serve as a model for how we might get in touch with Asher Lev's creative ambiguity, how we might find diversity within unity, many points of view that are all valid, many truths that are all "objective."

I think it would be useful to recall the story of Schrödinger's cat and his place in the underlying nature of physical reality. This cat, remember, is a quantum cat who has been placed inside an opaque box with the equal likelihood that he will eat some poison or some wholesome food. If he were an ordinary Newtonian cat, the kind that most of us live with, he would eat one or the other and as a result he would be either dead or alive. But because he is a quantum cat, he eats both the poison and the food, and he is both dead and alive.† He is, as a quantum physicist would say, in a superposition of both states. His wave function is spread out across both possibilities.

The cat's many-possibilitied quantum reality only "collapses"

---

*As presented in Heisenberg's interpretation of quantum theory.

†For a fuller account, see Chapter 2.

into a single everyday actuality (alive *or* dead) when we open the box and look at him. The act of looking at Schrödinger's cat is akin to what happens in the laboratory when a physicist measures a quantum system. At the moment of measurement, the physicist literally plucks one possibility out from many-possibilitied quantum reality and actualizes it. He acts, as I have said, as a kind of midwife to reality, drawing out one face of its potential and making it real in our everyday sense. I believe that our conscious minds are in the same relation to the many faces of truth.

I believe that in our conscious life we are all midwives to reality. We are a bridge between the realm of potentiality and the world of actuality. Through our imaginations we are in touch with a plethora of possibility, yet in our focused thoughts we must choose one from the many. When we focus our attention on a situation or a problem, we objectify it in just the way that we either kill or liberate Schrödinger's cat when we look at him. I am reminded of the passage in Stravinsky's autobiography where he says that whenever he sat down to compose a piece of music he had the uncomfortable realization that he had all the notes in the universe and every style of music ever created to choose from. Yet when he actually wrote his notes on the page, he chose one style, and one uniquely his own. The suggested physics of this is that the act of focusing, concentrating, pumps energy into the brain and collapses the wave function of thought. It transmutes many possibilities into one actuality.

But this objectification through focusing is not reality *creation,* as the postmodernists would claim. We do not *invent* reality through the perspective of thought. When we open the box on Schrödinger's cat, we do not see a dog or a rabbit. Stravinsky's music is not that of Wagner or Chopin. There is an underlying, catlike reality latent within the cat's wave function, just as there is an underlying Stravinsky-like pattern latent in the pool of potential from which he drew his music. Similarly, though all of the people we know can draw on a rich vein of possible expression or behavior, most display a certain inexpressible "style" or behave "in character." In all these examples, there is something *there,* something real for us to abstract from. When we pluck one aspect of that reality from its vast sea of potential,* we are really creatively *discovering* it. We are real-izing its own potential.

Quantum reality shows us that there can be many points of view, or many faces of truth, some even mutually contradictory, and yet all equally real in the potential sense that all of the quantum realm has existence. It is equally real that Schrödinger's cat is alive and that he is dead. Both are potential truths reflecting the deeper, underlying reality.

In the same way, Christianity, Judaism, and Buddhism may be equally true and equally real even if mutually contradictory. It is possible, as Asher Lev notes about some art, that some things may be both "creations of the Master of the Universe . . . and monstrous birthings from the realm of darkness," that we may sit in rooms "that are weighty with the possibilities of reverence and corruption *simultaneously*."[17] All are *aspects* of reality, objectifications of a deeper underlying truth that contains them all as possibilities.

In putting together our spiritual or conceptual systems, we behave like the physicist who measures the quantum event. We act as midwives to reality through our objectifications of that reality. We pluck out one of an infinite number of possibilities and actualize it. Once it is actualized, it can be measured and observed and any two or ten observers will agree on its properties. It is as objective as anything else that science has ever studied. But this way of looking at objectivity differs radically both from postmodernist relativism and from the monolithic truth of modernity or religious absolutism.

Theater director Peter Brook has found a very direct and powerful way to illustrate the creative dialogue between one's own expressed potentiality and the pool of deeper, underlying, and multifaceted potentiality through the medium of the dance. In one of his improvisations, he collected a group of dancers drawn from many countries and cultures, none of whom could communicate directly through language. He found that while each dancer had the clear impulse to express his or her own cultural style (cultural

---

*The cat is really in an infinity of superimposed states, not just the simple two of aliveness and deadness that we have discussed.

truth), it was also true that all were able to get beyond that individual style to something that could only be discovered and expressed through their common dance:

> Everyone can respond to the music and dances of many races other than his own. Equally one can discover in oneself the impulses behind these unfamiliar movements and sounds and so make them one's own. Man is more than what his culture defines; cultural habits go far deeper than the clothes he wears, but they are still only garments to which an unknown life gives body. Each culture expresses a different portion of the inner atlas: the complete human truth is global, and the theatre is the place in which the jigsaw can be pieced together.[18]

It is the ideal of a quantum society that our shared public life, too, could be a place where that jigsaw is pieced together, a place where "the inner atlas of complete human truth" might be creatively discovered. It is true, as the deconstructive postmodernist says, that there is no platform on which to stand *outside* the system of our observations and interpretations, outside our perspectives. We are always *inside* our dialogue with underlying reality. The observer is part of what he observes. But it is not true that we have *only* our interpretations and perspectives. We are always members of that larger dance.

It is not true that "truth" is only our relative point of view, and that one "truth" is as good as any other. There are criteria. There is an underlying reality there. The cat can only be some sort of cat. We are in dialogue with *something*. It is just that that something has many faces, many potentialities, and that the more of those faces we can know the closer we will come to being in touch with the larger underlying reality. Thus, our inner or shared life is a dialogue between the clearly defined foreground and the ambiguous background, definite contents and unspoken understandings, verbal analysis and nonverbal images and gestures. The presence of *both* allows them to evolve creatively. Where the deconstructive postmodernist is a dilettante, tasting a little of this and a little of that, the citizen of a quantum society seeks seriously to become multilingual, at home with many faces of our human potential for language.

The truth of quantum reality is an ambiguous truth. It calls upon us to live with the possibility of other possibilities. But such ambiguity can be creative. In the physical system in the brain out of which consciousness arises it *is* creative. Self-organizing quantum systems thrive on ambiguity. They are, as Ilya Prigogine has pointed out, delicately poised between order and chaos. They exist at what he calls "far from equilibrium conditions."[19] If a self-organizing system becomes too static, it runs down; if it becomes too chaotic it breaks apart. It needs a creative balance of order and chaos to thrive. This is ambiguity.

For both the absolutist and the relativist, ambiguity means paralysis. Neither sees the possibility of commitment in the absence of certainty. As Asher Lev's father says, "From ambiguity I would not derive the strength to do all the things that I must do." That same absence of certainty condemns the relativist to the status of Richard Rorty's dilettante. He will familiarize himself with all the "language-games" but commit himself to none. There is no *reason* for such commitment.

Isaiah Berlin has written, "Principles are not less sacred because their duration cannot be guaranteed. Indeed, the very desire for guarantees that one's values are eternal and secure in some objective heaven is perhaps only a craving for the certainties of childhood or the absolute values of our primitive past."[20]

Both the multifaceted nature of quantum reality and the limitations imposed by Gödel's Theorem call upon us to accept the partial nature of the truths to which we can have access. But both make us keenly aware that these partial truths are but the obtainable faces of a deeper, underlying, though ultimately inexpressible, truth. Such awareness requires the more mature attitude toward our commitments suggested by Isaiah Berlin's words. This is a maturity that allows me to be a committed Christian or a committed Jew, a committed advocate of my own vision of the good or the true or the beautiful, *while at the same time* allowing me to acknowledge that my way is only one possible way.

An analogy that may help to bring out the possibility for a deep commitment to a partial truth is my relationship to my husband.

**FIG. 6.1 EACH ACTUALITY IS A *PARTIAL* EXPRESSION OF AN UNDERLYING, MORE ALL-EMBRACING WELL OF POTENTIAL.**

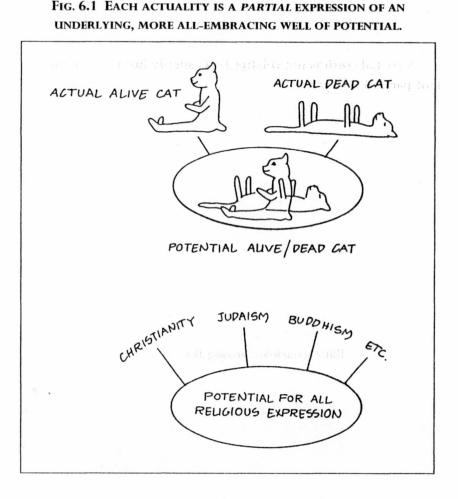

My commitment to him is for life and without qualification, yet it is made with the knowledge that other women have similar commitments to other men. It is even made in the knowledge that I might well have married some other man, that this is only one *possible* marriage for me.

The maturity that I must cultivate is a maturity that stands in the face of ambiguity and that tolerates the vision of others, a maturity that allows me to live in a complex and pluralistic society without losing my bearings. "To realize the relative validity of one's convictions and yet stand for them unflinchingly," says Berlin, "is what distinguishes a civilised man from a barbarian."[21] I would change these words only to replace "relative validity" with "partial validity": To realize the *partial* validity of one's convictions and yet stand for them unflinchingly is what distinguishes a civilized man or woman from a barbarian.

A partial truth is not relative. It is simply just part, an important part, of a larger story. A mature commitment always, in the end, includes a commitment to that larger story, or to its underlying potential (Fig. 6.1). In the case of my marriage, I am aware that part of my commitment is to the institution of marriage itself, to the underlying potential of the married state, and thus even to a concern for the enduring quality of others' marriages. This takes me beyond the limiting particularity of my own situation while at the same time leaving me deeply rooted within it. The particular (my own marriage, my own religion, my own set of deeply held values) is my necessary path to communion with the deeper, more all-embracing reality. As T. S. Eliot expresses it,

> Below, the boarhound and the boar
> Pursue their pattern as before
> But reconciled among the stars.[22]

# 7. A Universal Evolutionary Principle

*When every certainty is shaken and every utterance fails, when every principle seems doubtful, then there is one ultimate belief that can guide our inner life: the belief in an absolute direction of growth to which our duty and our happiness demands we should conform.*

Pierre Teilhard de Chardin*

We left off our discussion of truth with the suggestion that all particular truths are partial, each an aspect of some deeper, underlying potential. But for many of us that may not be sufficient. Human beings are not usually content with partialities and halfway houses.

Once we become aware that there is a deeper truth presently beyond our grasp, we feel a relentless urge to know it, to plumb deeper until we get to the "bottom of things." This urge makes our intelligence inclusive. Anything that we come across we struggle to include within our general view, we struggle to integrate it into our way of life or our system of thought. This is how human beings evolve as cultural beings.

For the deconstructive postmodernist, of course, there is no evolution. His sense of the "used-up-

*Cosmic Life.*

159

ness'' of all form leaves him, as Francis Fukuyama put it, foraging about in the museum of history. He is destined always to relive only what has been. So his partial, or relative, truths take him nowhere. His doubts and his uncertainties are for nothing. There can be no further progress. All experiment, all trials, all particularities are just so many elements of play. We can dabble, like Richard Rorty's philosopher dilettante, but we can never arrive at any further conclusions.

In a curious way, the deconstructive postmodern rejection of progress harks back to the preevolutionary worldviews of Plato and the Christian Church. In both Plato's theory of the Forms and in Christianity's emphasis on the Creation story and the immortal soul, there is a sense that everything interesting that *can* happen already *has* happened. The Forms contain all possible knowledge, the Creation contains all possible existence. The result of both is a static view of the human. Nothing new or interesting can ever happen. Everything that can be is.

Both Platonism and Christianity, in consequence, had to distance human beings from their surrounding environment. Nature and experience were seen as the world of disturbing change, a world of error or chaos. Doubt and uncertainty, for the Christian, are the work of the Devil.

On a quantum view, by contrast, both the role of experience and the possibility of progress are radically different. Quantum persons are thrown headlong into experience. The observer is part of what he or she observes. And there is always more to observe, more to discover, more to real-ize. Quantum reality (the Schrödinger wave function) has an infinite potential for unfolding. It is essentially evolutionary. Human consciousness too, I suggest, because of its likely origins in quantum dynamics, is therefore essentially evolutionary.

So I believe that the basic process of cultural evolution is both real and unavoidable. An evolutionary view is necessary to our pursuit of truth, or to our pursuit of a ''higher'' or a ''deeper'' or a better life. It moves us always beyond the given, the partial, the status quo. The multiple, the uncertain, and the ambiguous present themselves as opportunities for further discovery. But change or movement for their own sake are not always creative.

The adoption of just any idea does not always lead to a further or a deeper truth. Some ideas are clearly bad. Some "truths" are not true at all.

So how, in a pluralistic world, do we decide? How do we know whether the changes and the multiplicities with which we are confronted are following what Teilhard de Chardin calls "an absolute direction of growth to which our duty and our happiness demands we should conform"? Or whether they are taking us down the path of error? How do we sort out the ephemeral or the trendy from the lasting or the "absolute"? How do we know a good idea from a bad one?

I think that to answer these questions we must look at the process of evolution itself, and see how the changes affecting us and our social and cultural development fit into that process. Is there, as Teilhard and others have believed, one absolute pattern of growth in the universe in which we can see all growth, including our own, reflected? Can we, as I suggested at the beginning of this book, find laws and principles of nature that are the same as the underlying laws and principles of our own development? And if so, can we use a knowledge of this deeper, universal process to guide our own rough ride through the complexities and the ambiguities of the new society into which we are moving?

## Natural and Cosmic Selection

The word *evolution* itself implies progress. It implies change for the better, change in a direction. So if we want to know whether any particular change is an evolutionary change, whether any idea or lifestyle is good to adopt, we must ask where it is taking us, what it achieves. But for that we need criteria.

All evolution, at whatever level of reality it occurs, follows the same basic pattern first noted by Darwin in biology. That pattern is variation, leading to selection, leading to further variation. We see this, for instance, in trial and error learning. A baby makes all kinds of sounds before it learns to select those sounds useful to its own language. A rat pushes several levers until it learns that a particular lever will release food. When attacked by a viral infection like measles, the immune system produces a flood of antibodies, from which one is selected for its effectiveness in combating the disease. And cultures, too, throw up many different trends, most of which are short-lived, until a few are selected to carry on.

In asking what direction an evolutionary change is taking, we are really asking what are its principles, its criteria, of selection. How is one possibility chosen from among the many offered by variation? What qualities are selected for?

Darwinian evolution itself has a very clear set of criteria for what determines evolutionary selection. Out of the myriad possibilities thrown up through random biological mutation, those mutations that lead to a more successful adaptation to the demands of the environment will be selected. That is, the choice of one mutation over another as the "better" mutation is determined through natural selection for adaptive capacity. According to Darwinian theory, those mutations thus chosen through natural selection will survive, while others will die out. Natural selection for survival, we are told, is nature's criterion for successful change.

Darwin's work, as we know, was revolutionary in its time and it remains until today the bedrock of orthodox biology. In its larger claims it now seems unquestionable. Life clearly did begin as very simple organisms that slowly evolved to more complex ones. Some of those organisms seem more clearly adapted for survival than others, and a principle of natural selection seems to work in identifying which these are. But there are many ways in which Darwinian theory itself is terribly parochial. It raises as many questions as it answers.

Does evolution happen, for instance, in exactly the way that Darwin argued, and only in that way? Are the mutations that give rise to variation themselves random, and are the criteria for selection necessarily determined by "blind" adaptation? Can we, through the evolutionary process, ever rise above our specific environment?

The case of the gene that causes sickle-cell anemia illustrates the extent to which the Darwinian process itself is localized in environment. This gene is found almost exclusively in members of the black race. If only one gene is found in the blood, it protects the carrier against malaria, but if two genes are present they lead to a possibly fatal variety of anemia. Within the African environment, where malaria is a common threat, there is a clear evolutionary advantage to carrying the gene. But for black people living in northern Europe or America, where malaria is uncommon, the gene's protective qualities are clearly outweighed by the threat of anemia. Outside the African environment, the gene confers an evolutionary disadvantage.

It seems to me that Darwin's criteria for selection are too limited to give his account of evolution any clear direction. Its processes are random, locally defined, and wholly nonpurposive. It is, to use Richard Dawkins's famous phrase, the work of a "blind watchmaker." As it is environment dependent, Darwinian selection can never have any "absolute direction," as Teilhard would express it. We can't say, for instance, that it is a "good" or a "bad" thing to carry the sickle-cell gene. It just depends on where the carrier is living.

Darwin himself was an optimist, saying that "As natural selection works solely by and for the good of each being, all corporeal and mental endowments will tend to progress towards perfection."[1] But on the basis of his own theory he has no grounds for saying this. All that an orthodox Darwinian can say is that, given a specific environment, evolution will select those species best adapted to it. A different environment would select for different variations. In its original form, the theory is relativist.

This relativism explains why, when Darwinian evolution was applied to social progress, the theory could become what the *Encyclopedia of Philosophy* calls "all things to all men." In the hands of the right, evolution seemed to justify "survival of the fittest" in a strict laissez-faire environment. It was used to support Adam Smith's free market. For the left, it implied that social engineering to alter the social environment would encourage the desired sort of adaptations in human behavior. In an environment of thieves one would adapt to become the best possible thief; if honesty led to reward and success, one would become honest.

If we apply Darwinian principles to society, we are really back with the postmodernists. All criteria are local and relative. As a character in a Nicholas Mosley novel says, "If there is no guiding principle in evolution, then why should one form of behavior be better than any other?''[2]

In fact, natural selection simply doesn't go far enough to make it applicable outside biology. It has nothing to say about how life came about in the first place, and it was never intended as a commentary on society. Even within biology many contemporary evolutionists think it fails to explain the kinds and degrees of complexity that we find in nature. If blind adaptation were the sole criterion of evolutionary success, simple bacteria would be the most successful life form on earth. There is no strictly *scientific* Darwinian reason why life should ever have evolved beyond them. Also, according to Darwin's own criteria, the fact that the species now alive on earth, which are only 1 percent of all the species that have lived in the past, have temporarily won a local lottery, confers no special dignity or value on them. There is no sense that they are part of any overall progression. Everything is just random, or accidental.

Yet despite its inadequacies, adaptation to the environment has become the guiding ethos of most sociology and much psychotherapy (certainly that of Erich Fromm and Karen Horney). Both concern themselves with ironing out deviance, with the adaptation of the misfit to the demands of his or her social environment. For many people this may be necessary, and adaptation itself can be an important contribution to growth. Adolescents need to adapt their often wild or unrealistic demands and dreams to the social realities in which they are growing up. Immigrants to a culture must adapt to their new surroundings. Anyone breaking into a new situation or group must learn to adapt. But for large classes of people—creative people, people living in an emotionally harmful or unnourishing environment or, as Jung argued, people in the second half of life—mere adaptation may be destructive, boring, or even suffocating. Real growth often requires a balance between adaptation and dreaming, or between adaptation and rebellion.

The needs of many people require growth *beyond* the environment, growth toward some more ''spiritual'' (or intellectually, artistically, morally, or even materially better) goal. Cultural growth, too, if it is to happen, almost by definition must reach beyond the given. For these kinds of personal or cultural evolution, we would look for selection criteria that are in some sense more objective.

Very recently, an American astrophysicist has given Darwin's own evolutionary thinking what may be an important cosmic twist. Lee Smolin, at New York's Syracuse University, suggests that the existence of our universe, with its particular set of physical constants, laws, and principles, is in fact the outcome of a long chain of natural selection from among countless and varied parent universes.[3] His reasons for thinking this not only make sense, for the first time, of how our universe came to be, but they also add an important sense that there may after all be an absolute direction of growth operating at least at the level of cosmic history.

The universe as we know it today has clearly evolved since the Big Bang. Before the Big Bang, there was neither space nor time nor matter. And the universe has its own unique set of characteristics—the relative strengths of the four fundamental forces (strong and weak nuclear, electromagnetic, and gravitational), the masses of the various elementary particles, the fact that entropy is increasing from a low-entropy state, the fact that it has three spatial dimensions, and so on.

If any one of these, or a whole host of other characteristics, had been only *slightly* different, the universe couldn't exist as it does, and we couldn't be here. A whole, long chain of fortunate factors means that life is *just* possible. To support the existence of stars and planets and life, the universe also has to be of a certain minimum age. The Big Bang happened about fifteen billion years ago, the sun was formed about ten billion years ago, our earth is about four billion years old, and life itself began some two or three billion years ago. But if gravity had been only slightly stronger, the whole process of unfolding creation would have collapsed long ago. Life wouldn't have had time to evolve. If gravity had been only slightly weaker, the primordial gases couldn't have condensed into stars and planets in the first place. There are many other, similar, physical "coincidences" contributing to the fact that we are here.

The question that has plagued cosmologists for years is whether the universe's present delicate balance is just a very remarkable accident, just the random outcome of random cosmic events, or whether some noncoincidental process is at large that influenced it to be as it is. Smolin's work leads clearly to the conclusion that something more than mere coincidence is at work. It suggests the universe is the *expected* outcome of a particular series of cosmic selection procedures beginning with black holes and arriving eventually at the possibility of us.

Smolin's idea of cosmic selection at work takes its starting point from the argument that our own universe (one among many) is "closed," as though it were a black hole. That is, the universe is not infinite in either its spatial dimensions or its temporal duration. It has a beginning and an end. It has a surface, like a balloon, that moves out from some point of expansion (the Big Bang) and will one day move inward toward a point of contraction (the Big Crunch).

A black hole is formed when a heavy star burns up all its fuel, and thus loses the pressure from radiation that kept its gases fanning outward. According to General Relativity, a star must be at least three to six times heavier than our sun to collapse in this extreme way. A lesser collapse would lead to a white dwarf or to a neutron star.

Once it does collapse, a black hole also collapses all the space around it. Space is curved by gravity, and the gravitational field of a black hole approaches infinity. The black hole acts like a bubble drawing its surface inward. It forms its own, interior space-time, and "drops out" of the universe in which it was formed.

So if our universe *is* a closed black hole, it has a history, and perhaps a future. It not only began as a condensed singularity (collapsed space-time and matter) that "exploded," but one day will collapse back into another singularity. At that point, many cosmologists believe, the cycle of creation will begin all over again—the singularity will "bounce" back into another long period of expansion, creating an entirely new universe.

The truly exciting thing about the cycle of collapse and "bounce back" is that many cosmologists believe it does not happen only at the death of a universe. Rather, they believe, black holes are formed continuously inside existing universes, each one expanding to create a new universe. So the cosmic scenario is one of universes within universes, each with its own space-time and each the offspring of a parent universe.

Smolin's contribution to all this is twofold. Each time a new universe is formed by the expansion of a black hole, he suggests, it will have a slightly different set of physical constants—different gravitational strength, different forces binding the atoms together, different relative masses of the particles that make up atoms, and so on. And universes with different physical constants, he argues, have a varied chance of continuing the cycle of creation, just as plants and animals with different physical characteristics have a varied chance of continuing in the process of biological evolution. Some physical constants are more likely to lead to the creation of further black holes (and hence to more universes), and some will produce fewer. Those universes with more black holes will be more successful parents, have more offspring, and thus be at an evolutionary advantage.

Smolin's big idea is that a given universe's capacity to make stars decides how successful it will be in the ongoing drama of cosmic evolution. This is because stars are necessary to make black holes. Our own universe, as we know, has many stars, and it does so only because its physical constants are exactly those required to synthesize carbon atoms, which encourage the formation of stars from gas clouds. A different sort of universe, with different physical constants, might have fewer carbon atoms, thus fewer stars, and consequently fewer black holes.

So, by a delicately balanced set of selection criteria, our particular sort of universe has been chosen to exist because it has all the features necessary to continue the cycle of creation. Having these features may make no difference to the life of our own universe—it might exist whether it were a part of a larger cycle of creation or not—but it very probably makes an enormous difference to the likelihood of its containing the right elements for life and consciousness, as we shall see. Having the right universal constants to be part of a larger cycle of cosmic evolution also makes our universe seem less a freak (less improbable) and more something to be expected. The selection for a universe with these constants is absolute, not relative. The criteria of selection are the necessary, and the only, criteria that could lead to a continuation of the star-making, black-hole-forming, universe-creating cycle.

At the same time, Smolin's theory of cosmic evolution takes us to the very doorstep of biological evolution. The same carbon atoms necessary to star formation are, as we know, necessary to form other heavier elements and to sustain the creation of life. The planets that revolve around stars give us a place where life can live. So the same process of selection that chooses among universes leads to a further selection of our universe as one that contains the necessary preconditions for life as we know it. From the moment of star creation to selection for the main ingredient of the first organic molecule, we can see a single process of evolution at work.

A good theory of cosmological evolution like Smolin's also takes us halfway to grounding the Anthropic Principle, the set of ideas that concerns itself with how conscious creatures such as ourselves have come to exist in the universe. There are various versions of the Anthropic Principle, but we need only concern ourselves with the differences between the so-called Weak Anthropic Principle and the Strong Anthropic Principle.

All versions of the Anthropic Principle note that the universe in which we live seems to favor life and consciousness. The Weak Anthropic Principle simply notes this as a fact. It says something like: Out of all the possible values the physical constants of the universe might have had, it seems, improbably, to have arrived at that narrow range of constants that favor the existence of carbon-based life forms.[4] Some people have accused the Weak Principle of being a tautology: Because we are here, it must be possible that we can be here.

The Strong Anthropic Principle goes much further. It claims more necessity for the presence of people like us. It says something like: "The Universe must have those properties which allow life to develop in it at some stage in its history."[5] The Strong Principle claims that life is more than a coincidental fact. It is a fact that requires explanation.

A philosophical analogy of the Weak and Strong principles imagines a man confronted by a firing squad. All members of the firing squad take careful aim and fire their guns, but the man remains standing there untouched. On the Weak Principle, he shouldn't be surprised that he is still alive, because if he weren't alive he wouldn't be standing there to notice the fact. On the Strong Principle, he should indeed be surprised that something so against the odds should happen, and he would be right to ask for an explanation.[6]

Those who follow the Strong Anthropic Principle usually give one of three explanations for the presence of life and consciousness in the universe. First, there is the religious explanation. This explanation argues that there is a God who wanted there to be life and consciousness and so He built these into the structure of the universe. Second, there is the "many-worlds" hypothesis. Many-worlds theory argues that *every* possible sort of universe exists, so in at least one of these possible universes we would expect to find life and consciousness. And third, there is the evolutionary account, which argues that the universe is evolving in such a way that life and consciousness would be favored to occur at some point.[7] It is toward supporting this third explanation that Smolin's theory of cosmic evolution comes very close.

Smolin gets far enough to show why the universe would evolve in such a way as to favor the presence of carbon atoms, heavier elements, planets, and stars; i.e., black holes are necessary to make new universes, stars are necessary to make black holes, and carbon atoms are necessary to make stars. But his theory does not take us all the way to satisfying the Strong Anthropic Principle. Carbon atoms, and the further chemical elements they in turn produce, are still a long way from life. Smolin's work on its own doesn't show that life and consciousness are a necessary outcome of the evolution that results in carbon atoms. Even given stars and planets and the right chemical elements, the formation of life still seems to be very much against the odds. So where does life come from? Why *are* we here?

At this stage in the chain of creation, does mere accident finally take over? Is life, and hence consciousness, just a freak result of random events once the right chemical elements exist? Does this accidental life then, once formed, follow its own Darwinian evolutionary principle unrelated to the evolution that brought about its preconditions? Or is there, after all, some further extension of just one evolutionary principle operating both among the stars and among ourselves?

Many philosophers of the nineteenth and twentieth centuries have argued that there is just one evolutionary principle acting throughout creation. I think particularly of the "process philosophers" Bergson,[8] Whitehead,[9] and Teilhard.[10] All have believed that consciousness was immanent in the structure of the universe from the very beginning, a natural part of reality, that it has evolved slowly to human consciousness, and that this evolution will continue—either to an "Omega point," as for Teilhard, or indefinitely, according to Bergson.

The recent impact of cosmology and theories of cosmological evolution like Smolin's has given process thought a new impetus among both philosophers and theologians. But if the various evolutionary life forces of the process philosophers are to make any scientific sense, they must be grounded in the mechanism of some experimentally testable, and *physical,* evolutionary process. They, or their scientific equivalent, must be seen to play some active part in the actual physics of the universe, that same physics that gives us stars and planets and carbon atoms.

I want to return to the question of selection criteria, the selection criteria used in any evolutionary process. At the moment, physicists working at the level of fundamental quantum physics do not know what these are. Quantum theory, as we know, says there is a wave function that carries all possibilities. Everything that is has a wave function. Even the universe as a whole has a wave function. But the transformation from wave function to reality, from multiple possibility to one actuality, is a mystery. No one understands how, or why, the wave function "collapses." This is perhaps the single most outstanding problem in physics today, yet its dynamics, which must involve some sort of selection principle, could well represent a kind of evolution at the level of fundamental, physical reality.

We live in a quantum universe. Everything—the stars and planets, the carbon atoms, life, and if it is part of the physical world, consciousness itself—obeys the laws of quantum mechanics. So it is here in quantum mechanics that we must look to see if there is a universal evolutionary selection principle at large. In particular, we must ask whether there is any possible solution to the collapse of the wave function that favors the selection of life and consciousness as things existing in this world.

There are, as the physicist John Bell noted, six main kinds of theories as to why or how the wave function collapses.[11] One, the Bohr theory, just says we can't discuss it; two of them, "many-worlds" and David Bohm's "hidden variables" theory, say it doesn't collapse at all; and another, the famous Wigner Interpretation, says that a disembodied, nonphysical conscious mind collapses it. Of these possibilities, only the Bohr theory is taken seriously by the vast majority of physicists, and it gives consciousness no place in the evolution of the universe. For Bohr, the observer (conscious or otherwise) is not represented in the equations of physics.

None of these various theories treats the collapse of the wave function as an actual *physical* process due to a specific set of physical circumstances (rather than to some metaphysical or extraphysical agent). Two that do are the "GRW," a theory that argues the wave function just collapses spontaneously,[12] and Roger Penrose's theory, which argues that it collapses due to gravity.[13]

I want to suggest a different physical theory, one that sees consciousness both as a part of the physical world and as an active agent in that world.* This theory has the advantage of introducing a meaningful selection principle for why the wave function collapses in a certain direction as it evolves from multiple possibility to fixed actuality. I believe this principle can show that evolution, even at the most primary quantum level where we are speaking simply of the evolution of the wave function, has an interesting direction that necessarily favors the eventual, and perhaps the permanent, presence of life and consciousness in the universe.†

---

*In the technical language of philosophy, such a theory would be both "monist" (nondualist) and "nonepiphenomenalist."

†The theory of wave function collapse that I am about to outline is the work of I. N. Marshall, this book's coauthor. A layperson's account appears in the published *Proceedings of the 2nd International Symposium on Science and Consciousness.* See I. N. Marshall, 1992, and also I. N. Marshall, 1993, for a formal, mathematical treatment.

## *Consciousness and Evolution*

If we want to ask what place consciousness occupies in the physical world and how it can be a consideration in the evolution of that world, we must return to the physics of consciousness. That is, we must consider the nature of the physical system in the brain that may give rise to conscious mind. Where would that kind of physical system fit into the rest of the physical universe?

We saw earlier that there are very good reasons for thinking consciousness arises from a quantum physical system in the brain that has the properties of a Bose-Einstein condensate. Bose-Einstein condensates are the most coherent, the most unified quantum structures possible in the physical world. In the case of the brain, such a Bose-Einstein condensate might emerge from the synchronous oscillations of neuron modules across the brain. It would be an emergent wave phenomenon.*

Wherever we have vibrations and waves, we have phases (the stages of development the waves have got to as they move along—like the phases of the moon or, more metaphorically, "the phases of one's life"), and patterns of phase difference (how the phases of different waves or developments are synchronized—like the time lag that separates the cycle of increasing hours of sunlight and the cycle of increasing temperature in the spring or the similar time lag between the entry of different instruments into the playing of an orchestral piece). Phases and phase differences are about timing and coordination and are unaffected by the sizes or the shapes of the waves being coordinated. The timing of an orchestra is unaffected by how loudly or softly the instruments play. In any Bose-Einstein condensate, these phase differences represent "excitations," or "patterns" on the surface of the condensate. They are like ripples on a pond.

In a Bose-Einstein condensate like a laser beam, the excitations due to phase differences represent information written on the laser beam—e.g., a hologram. A hologram is nothing but a pattern of phase differences between its underlying waves. In the physics of the mind, such excitations, or phase differences, could account for

*See Chapter 2 of this book, or Chapter 6 of *The Quantum Self*, Zohar, 1990.

the presence of mental information—i.e., specific thoughts or sensations or experiences.

What I am suggesting here is that when the quantum wave function collapses, it collapses *toward* definite values of phase difference.* This is like a chamber orchestra playing in time, with a definite beat, rather than in an uncoordinated way, or like a group of soldiers marching in definite step rather than raggedly. Since phase differences are sharper in Bose-Einstein condensates than in most other physical structures, this means that collapse of the wave function would *favor* the formation of Bose-Einstein condensates or the formation of definite excitations on Bose-Einstein condensates (an increase of information). That is, collapse would *select* for such formation. The operation of such a selection principle is compatible with existing laboratory observations, as we shall see in a moment.

Since, according to Fröhlich's and Del Giudice's work (Chapter 2), all living systems contain Bose-Einstein condensates and, according to my† extension of this work, consciousness, too, very likely arises from a Bose-Einstein condensate, the collapse of the wave function toward Bose-Einstein condensates means a collapse toward life and consciousness. This has very exciting implications.

If my suggestion holds up, we have found an evolutionary selection principle at the heart of fundamental physics that would favor the development of life and consciousness in the universe. The quantum wave function itself, I want to suggest, in its tendency to collapse toward states of definite phase difference, selects for those physical structures that are in fact the basis of life and consciousness.

There are some important pieces of reasoning or evidence that support the notion that the wave function does in fact use collapse toward some definite phase difference as its primary selection principle.

First, a wave function using this selection principle for collapse should also (according to the mathematical proof) collapse toward

---

*I.e., toward "eigenstates of a phase difference operator." See I. N. Marshall, 1993, for the mathematics of this.

†My own views are based on I. N. Marshall's work. See I. N. Marshall, 1989.

things that are very similar to Bose-Einstein condensates—crystals (which includes large structures like metals)* and electric currents running through metals. These things, conscious observers and macroscopic measuring devices comprising crystals and electric currents, are exactly what we would expect to see collapsing the wave function given that measurement (with photographic plates and electric devices or the human eye) is known to do so. Thus the theory is in keeping with known experimental evidence. It predicts the right outcomes.

Second, an evolutionary principle at the level of fundamental physics that selects for life and consciousness supplies the missing link between Smolin's theory of cosmic evolution and the eventual appearance of creatures such as ourselves in the universe. Given a universe that contains the right chemical elements, it would explain why these elements are more likely to be formed into Bose-Einstein condensates. Thus we can see one fundamental evolutionary selection principle at work in both the creation of our universe and the creation of life within it. This is a physical explanation for the Strong Anthropic Principle.

And finally, if the wave function collapses toward positions of phase difference, and thus favors the evolution of life and consciousness, we can see it as a solution to a great enigma within biological evolution itself. This enigma, the problem of "adaptive evolution," has startled the world of biology in the past few years and shown that Darwinian principles of selection on their own are inadequate to explain several recently noticed events.

"Adaptive evolution" was first noticed in experiments done on bacteria by Harvard's John Cairns.[14] Cairns discovered that, when their survival is threatened, bacteria can mutate many times faster† than the rate predicted by Darwin. In a recent experiment done by a fellow biologist,[15] the subject was mutated bacteria that could not produce the essential nutrient tryptophane. When these bacteria were grown on dishes that contained only minute quan-

---

*Crystals have modes of vibration, each one of which is in a definite state of phase difference, but the resultant energy is not concentrated into a single mode, as in a Bose-Einstein condensate.
†As we shall see in a moment, a subsequent experimenter has found this to be one hundred million times faster.

tities of tryptophane, most died, but after only twenty days some colonies began to grow again (according to Darwin this should have taken vastly longer). These survivors had repaired the damaged genes that could not produce tryptophane, but they showed no other mutations. If all their genes had mutated at this alarming rate, the bacteria would have been destroyed—unless natural selection itself had operated one hundred million times faster than normal to choose only those mutations that would lead to survival. On the basis of conventional evolutionary principles, this is impossible.

There is no known explanation for "adaptive evolution" in biology itself, but the presence of an underlying, or complementary, evolutionary principle selecting for life (where the bacteria's survival is threatened) at the level of fundamental physics could provide one. The many branches of the bacteria's wave function would be expected to collapse preferentially to one of the few that contained a future Bose-Einstein condensate, i.e., a live bacterium. This selection at the physical level would only happen when the bacteria's survival is threatened. If natural selection happening for survival is going on at two different levels, the one biological and the other physical, its rate could be very much faster than that predicted for biological evolution alone. What's more, this underlying evolution at the level of physics would introduce an absolute direction into biological natural selection itself.

Darwinian evolution on its own, we saw, is relative. Those organisms should be selected for that are best adapted to whatever environment in which they find themselves. Given this principle alone, bacteria are the most adaptive creatures in the world, and there seems no reason for evolution ever to have got beyond them. There is no reason for a selection toward complexity in Darwin's theory. But in an evolutionary principle selecting for ever greater life and consciousness, greater complexity would be expected to result. Higher, or more complex, life forms would have more complex patterns of phase difference on their Bose-Einstein condensates—larger ripples on their larger "ponds," as it were. These would be favored in collapse of the wave function. The enormously complex patterns produced by the human brain would be the most favored of all.

There is, then, at the level of the most fundamental physics, strong suggestive argument for a single evolutionary principle that selects for life and consciousness at every stage of the universe's unfolding. This principle is that the collapse of the quantum wave function itself is biased toward selecting for those physical structures (definite values of phase difference) that involve the type of organization required for life and consciousness. This same selection principle can account for how our own universe with its stars and planets and carbon atoms evolved from among many varied parent universes,* for why life itself arose from the primordial "slime" existing at the early dawn of the planet, and for why life has evolved to produce complex, conscious creatures such as ourselves. I believe that this same "absolute direction of growth," as Teilhard would call it, can be seen to be at work in the further development of consciousness and all its constructs. This, as I shall argue in subsequent chapters, has large implications for the evolution of society. This is not to say that evolution need be rapid or certain. Individuals can regress and whole civilizations can collapse. As in the GRW theory, what I am suggesting is a rather weak influence, almost undetectable in individual cases but having a cumulative effect over large enough systems and long enough spans of time—perhaps millennia.

In the actual physical dynamics of consciousness I have suggested in this book, consciousness is a pattern-making process. It is one kind of self-organizing system. So through its physics, a quantum-based consciousness constantly would be forming new, ordered, coherent wholes—that is, new patterns that draw chaos into order, or that draw simpler patterns into more complex ones. It would be like a whirlpool drawing surrounding water molecules into itself, only in the case of consciousness the "molecules" being drawn in actually would be fragments of experience, or bits of information.

On a quantum interpretation, the evolution of consciousness, like the evolution of universes or the evolution of the biological gene pool, proceeds from variation, to selection, to further vari-

*To justify this particular step in the argument we would have to go into a discussion of "backwards causation" in quantum physics, which I feel is too technical for the discussion here.

ation. At the level of fundamental physics it proceeds from the many possibilities carried within the wave function, to the selection of one actuality on collapse, to a further fanning out into more possibilities. For consciousness, this would be a process leading from multiple fragments of experience, multiple choices, multiple "simpler" patterns of information to a selection for organized greater complexity (more information), and leading on eventually to further variation on that complexity. Each choice would entail further choices.

In the case of multiple truth, with which we began our discussion, it would be a process leading from an array of possible, or partial truths, to a selection for further truth that draws elements of the partial truths into a more complex pattern. The process is analogous to the increasing complexity of character description in literature since Homer. Homer's characterization wasn't "untrue," it was just simple. Dostoyevsky's characters aren't "more true," but they mark a clear evolution of the genre. Classical music, too, from plainsong (Gregorian chant) to Beethoven underwent the same sort of evolution, the same movement from the simple to the complex. Many strains in our culture during these past three and a half thousand years have followed this same pattern. In recent years, with the information explosion resulting from new technology, the process seems to have accelerated. We have got into a stage of accelerated complexity, and hence, perhaps, into a stage of accelerated evolution.

Thus, when we ask, "How can we choose?" or, "How do we proceed in the face of many partial truths?", the answer is that we seek a richer, a more complex pattern that embraces essential elements of the simpler variations. This is not to say that we should seek a vague syncretism, nor the nostalgic pastiche of the postmodernist. Evolution does not proceed by taking a little bit of this and a little bit of that and jumbling it all together. Rather, it is a growth *out* of a partial truth and a further growth beyond it.

I am reminded, as I say this, of Bartok's use of peasant melodies in his composition. He does not simply collect them and throw them all together. His music transmutes the peasant songs, it evolves beyond them (though out of them) toward something new, something more complex. All creativity grows out of possible variations on a given tradition, which it then transforms. Our partial truths are the variations on which further selection can feed, the simpler patterns out of which more complex ones evolve.

In picking our way through the maze of partial truths and apparent truths, a bad choice would be a choice leading away from the direction of evolution. It would lead away from the enrichment and growing complexity of consciousness—toward isolation rather than toward integration. The fundamentalist makes such a choice. He opts for the simple and the nonambiguous over the complex, the singular over the plural. He opts to isolate himself from possibility. He treats the given as a final certainty rather than seeing it as a jumping-off point for further and more complex reality.

Those who would work with evolution, by contrast, open themselves to the varieties of experience. They stand poised before the possibilities of experience, willing to let these unfold. This poise calls for a holding on while letting go. The Jew or the Christian remains firmly rooted in his tradition, deeply committed to it, while at the same time recognizing that it is not the only tradition. This recognition itself, if allowed to happen, will take the believer to a more complex position.

Whenever I recognize that my way is not the only way, I am already on the path to taking a further way. Whenever I am willing to admit there is more to understand, I *have* understood more. My consciousness, and the physical patterns out of which it is formed, has taken an evolutionary step.

Given the possible physical dynamics of consciousness I have outlined, given that the possible physics of the mind is bent on forming ever more complex, ordered coherent wholes (more complex patterns), such evolutionary steps are the norm to be expected. I describe the process of riding with evolution as a kind of ''letting go'' because it is a process of trust, a willingness not to defend against ambiguity and multiple possibility, a willingness not to grasp at simplicities or apparent final and exclusive truths. T. S. Eliot seems to be describing such a process in the second of his *Quartets:*

> Home is where one starts from. As we grow older
>   The world becomes stranger, the pattern more complicated. . . .

We must be still and still moving
Into another intensity
For a further union, a deeper communion. . . .[16]

Involvement in the process of evolution is, I believe, that deeper communion.

Evolution, as I have outlined it here, is a never-ending process. There is no "omega point" at which all partial truths will be gathered into one absolute and ultimate expression of Truth Itself—no "end of history," to use Francis Fukuyama's (or Hegel's) portentous term. There is no ultimate religious or social reality that is the end point of the evolutionary process. Truth itself, certainly for any finite being, *is* a process of unfolding, a process leading always to more possibilities. Ambiguity and multiplicity are the food on which it grows. But it is a process with a direction. It is a process with which we can align ourselves and in line with which "our happiness demands we should conform" (Teilhard). It is a process through which we realize ourselves, both as individuals and as groups and societies.

There are many concrete implications of there being one universal evolutionary process. It tells us that we live in a "friendly" universe of which we and our consciousness are fully a part. It tells us that there is no meaningful separation between ourselves and the physical and natural worlds. It tells us that we, and all the constructs of our consciousness, are part of a process of growing complexity. It tells us what role multiplicity and pluralism play in our lives. I want to turn now specifically to that pluralism. I want to look at pluralism and our social processes. I want to ask what role (and why) pluralism plays in the evolution of society, and in what social direction this evolution is taking us.

# 8. A Community of Communities:* Conflict and Tolerance in a Quantum Society

*We have to learn to speak to those we do not wish to convert, but with whom we wish to live. . . . In a society of plurality and change, there may be no detailed moral consensus that can be engraved on tablets of stone. But there can and must be a continuing conversation, joined by as many voices as possible, on what makes our society a collective enterprise; a community that embraces many communities.*

Rabbi Jonathan Sacks[†]

I recently attended an international symposium on science and consciousness in Athens.[1] The 150 invited participants came from twenty different countries and cultures and were from a wide range of academic or professional disciplines. Some were psychologists or linguists, others were physicists, biologists, philosophers, anthropologists, spiritual leaders, architects, or artists. Some were Hindus or Buddhists, others Christians, Jews, or Japanese Shintoists. Each spoke a private language associated with his or her own work, affiliations, or thought processes.

The symposium organizers hoped to end the

*I have taken my title from the sixth of Rabbi Jonathan Sacks's 1990 BBC Reith Lectures. Jonathan Sacks, 1991.
†*The Persistence of Faith*, pp. 64, 68.

week's meetings with a report that would reach certain conclusions and make a definite set of recommendations. Each day we were asked to discuss a set of given questions in small groups and then report our agreed reactions at the end of the day to the symposium as a whole. But nobody could agree on much of anything. At first we could not even find a common language in which to speak to each other. Individuals simply subjected others to their own preset opinions and hobby horses and were in turn shouted down, misunderstood, or ignored.

The symposium structure did not allow us to quit or go our separate ways. No one was allowed to dominate. Much to our surprise, after being together sixteen hours a day for five days, we eventually got to the point where each could listen to others. Indeed we listened eagerly and with empathy and grew through doing so. But we did not overcome our differences. No consensus of all views was reached, and participants felt that any attempt to express one would mask the rich variety of opinion and points of view that had emerged during our continuing conversation.

In the end, the idea of a monolithic symposium report was abandoned in favor of performing a dramatized version of a "Socratic dialogue." Each of the eight characters gathered in the modern Athenian dialogue expressed separately some different point of view, while the dialogue as a whole successfully represented the rich diversity within unity and fellow feeling that had come to characterize the symposium itself. The dialogue was our common reality.

I believe that this symposium, and its outcome in dialogue, is an apt symbol for the wider pluralistic societies in which we might hope to find ourselves participating. Though of course, as we shall see later, such dialogue in society at large must be embedded in practical social and political decisions.

## The Challenge of Pluralism

Diversity and a juxtaposition of peoples, cultures, traditions, political and religious differences, and lifestyles characterize today's society as never before. Global communication and mass transportation, shifting patterns of work, and the greater mobility demanded by modern life have shrunk both physical and psychological distance. All over the world we are witnessing what the historian Arnold Toynbee called a vast *Völkerwanderung*, a swirling movement and migration of individuals, peoples, and cultures in pursuit of a different or a better life.

In the past 150 years, the majority of people in Western countries have moved from small, monolithic villages and farming communities to large cities with mixed populations. These same cities are swollen with immigrants from other countries and cultures. From the Caribbean and the Indian subcontinent they have come to Britain, from Yugoslavia and Turkey to Germany, from the Maghreb to France, and from Mexico, Central America, and the Far East to an America already in ethnic turmoil. We have yet to see the full force of an expected exodus from the former Soviet Union.

Within given national societies, both new arrivals and older ethnic or cultural groups struggle to express their differences, their uniqueness, while being brought face to face with others doing the same. The many separate nations that together comprised the former Soviet Union now rise up separately to declare their own identity and historical validity. In the United States, racial, ethnic, and immigrant groups that once eagerly subsumed themselves in the "melting pot" now passionately declare their right to cultural self-expression. Women, homosexuals, senior citizens, and even sufferers from AIDS all cry out for separate, and equal, recognition of their differences. As one prominent political philosopher has noted, there is no identity in America today that is not a hyphenated identity. There are black Americans, Jewish Americans, Hispanic Americans, Native Americans, Wasp Americans, etc., but very few just plain Americans. More strongly still, British columnist William Rees-Mogg has suggested that "the United States is unwinding, strand by strand."[2] The same fear is expressed in Arthur Schlesinger, Jr.'s recent book on American multiculturalism, *The Disuniting of America*.[3]

To a lesser but still noticeable extent, the same strident differentiation is making itself felt throughout much of the world. There are more differences between people now living side by side, and people are more aware of their differences.

The old hegemony of dominant cultures is breaking down, often with a sense of shock and threat. This has happened recently in both France and Germany, where bullish right-wing nationalist movements have risen up against the influx of "outsiders." It is noticeable to some extent in Britain, where the new "multiculturalism" being urged on the population is seen by many as at best a mixed blessing. The same ambivalence characterizes Britain's closer ties with the new European community, and in America there is heated debate about the supposed threat to Western values and Western traditions posed by too much emphasis on other values and other traditions in education and public life. At street level, the sense of threat can be more physical, especially between black and white or between host majority and immigrant minority.

Where our reaction to the other is one of threat, conflict and intolerance are the inevitable result. We see this in the often violent confrontations between communities in our crowded cities. We see it in the hundreds of civil wars and tribal conflicts that have brought death and destruction to countless millions since the end of the Second World War alone. While I was writing this book, the United States experienced its worst race riots for decades, and parts of the city of Los Angeles were literally burning with interracial fear and hatred. In Yugoslavia, warring ethnic communities are killing each other.

I think it is not too strong to say that the challenge of pluralism, the challenge of learning to live with others who are different from ourselves and doing so in a creative way, is one of the greatest challenges facing social, national, and international stability in our times. It is a problem at least as great as those arising from the radically unequal distribution of wealth and opportunity within given societies and among different societies or nations. Indeed some strife supposed to arise from economic inequality often masks a deeper conflict of lifestyles or values. Pluralism strikes at the heart of identity, at the heart of who we are and what we value and how we look upon however little or much we have. It raises issues for which people will sacrifice their own lives or take the lives of others.

When we perceive the other as threat, the paradigm of modern defense analysis, there are three basic ways we can deal with him.

One strategy is an attempt to force him to surrender or to assimilate, to die or be enslaved or to adopt our ways. This is the path of conquest and domination. It is the path perceived by many blacks and Muslims in America and Britain. As an American black told the writer Studs Terkel,

> For some reason black culture is rejected in American society. It is looked upon with scorn by the majority. . . . There is a certain theft going on. In order for a culture to be accepted or co-opted, it's diluted. We're going to make it more acceptable by diluting it the way the majority thinks.[4]

Alternatively, if we perceive the other as threat, we can exclude him, build a boundary around ourselves and our culture and require that he stay outside it. This is the path of apartheid or segregation. It is the path chosen by those who would force minorities into ghettos or by those minorities who themselves choose to live in segregated communities, who choose to study in segregated schools or to do business only with their own "kind." It is the path of "black nationalism" in America and of extremist Islam in Britain.

Finally, if we perceive the other as threat we can treat him as different but equal, as someone with whom we will share a neutral public sphere while each pursuing our differentness, our otherness, in private spheres of individual social reality (in separate "atoms" of social reality). This is the path of liberal coexistence.

Liberal coexistence seeks, often successfully, to avoid confrontation and open conflict through a balance of mutual self-interests, but it can lead to isolation and fragmentation. In reality such coexistence is just a mutually agreed upon and more subtle form of apartheid. In a liberal society, the other remains wholly other and our peaceful coexistence depends upon our willingness to tolerate him, and he us. We agree each to live our own lives in our own way *beside* each other, we agree each to pursue our own ideals and our own dreams in private. We agree not to threaten each other, but inevitably, through modern communication and the natural effect of living beside others, even liberal coexistence carries its own sense of threat.

I am struck by the memory of my one-year-old daughter bursting into tears when presented with a huge pile of Christmas presents. The sheer extent of the choice offered overwhelmed her and caused her pain. The same kind of pain can result in society at large when individuals are exposed to a wide variety of cultures or lifestyles. In the absence of dialogue or some creative way to deal with the otherness of the other, mere choice can lead to fragmentation or to a relativization of one's own position. The one is destructive of social unity, the other of individual or group identity. As French philosopher Paul Ricoeur warns, "When we discover that there are several cultures instead of just one . . . we are threatened with destruction of our own by the discovery. Suddenly it becomes possible that there are just *others,* that we ourselves are an 'other' among others."[5] If true pluralism is to thrive and contribute to the evolution of society as a whole, both fragmentation and relativism must be avoided.

Conquest and domination, apartheid, and liberal coexistence are all mechanistic models for what might happen when different cultures, traditions, or lifestyles meet. They picture the self and others as something like a system of moving pendulums. Each pendulum (each culture) has its own characteristic rhythm of swing when left to itself, but this rhythm is influenced externally, through force, if it should get connected to another moving pendulum.

In conquest and domination, it is as though one of the pendulums is very heavy and it gets connected by a tight spring to a lighter one. For the most part, the lighter one takes on the rhythm of the heavier one. The supporter of apartheid sees the pendulums as more equal in weight and he would cut any connection between them, lest his own pendulum be thrown off its original trajectory. The liberal who believes in coexistence would ensure that any connecting springs between pendulums are extremely loose, so that only the most minimal influence is transferred as the pendulums swing in their separate trajectories (Fig. 8.1).

In all these mechanistic models the pendulums remain separate in themselves. They are related through force or influence, and the sharp image of the other is of one who might deflect me from my course. Both conquest and domination and apartheid or segregation deny diversity. Both would sacrifice diversity for the sake

**FIG. 8.1**

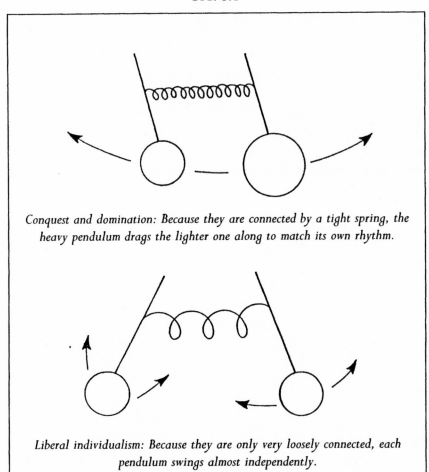

Conquest and domination: Because they are connected by a tight spring, the heavy pendulum drags the lighter one along to match its own rhythm.

Liberal individualism: Because they are only very loosely connected, each pendulum swings almost independently.

of an imposed or exclusive unity. Both deny the value of the other and respond to him with confrontation and intolerance. Liberal coexistence accepts diversity and recognizes the value, or at least the right, of the other in himself. But none of these classic responses to the other recognizes the value of his otherness to me. None has any model for diversity within unity, for a kind of creative dialogue in which differences meet and evolve through their meeting. None recognizes the value, or even the possibility, of a public "conversation" or a shared public reality.

I think, particularly, of the recent penchant in liberal societies for special-interest broadcasting. The ideal of those who support this is that each ethnic or religious group should have its own radio or television slot on which to express its views without concern for balance or broad appeal. Adherents can tune in and be reinforced in their convictions. Those who might disagree or who feel it is nothing to do with them can simply avoid listening and thus any possibility of offense. But there is no conversation here. As England's chief rabbi points out, this merely gives us "a situation in which opinion is ghettoised into segmented audiences. . . . We listen only to voices with which we agree."[6] Such ghettoization, he argues, is more appropriately called "narrowcasting." The real loser is any ideal of a shared culture, any ideal of differences or pluralism leading to shared growth or to real social evolution.

We have seen that any evolution, whether it is the evolution of stars, biological forms, or consciousness itself, feeds on diversity. The evolutionary pattern begins with variation, goes on to selection, and leads to a further variation. This is a universal process. We might expect, therefore, that the evolution of society, or of individuals within society, requires a similar variation of cultures or lifestyles, a variation of social possibilities from which a selection toward growth can be made. This is why complex cultures evolve more quickly than monolithic ones, and why larger cosmopolitan cities are more stimulating places in which to grow. But this variation cannot express itself only on "sidetracks." Evolution requires that the evolving system's many diverse possibilities meet. It requires the kind of "dialogue" that is missing in mechanistic models of the other and his otherness. Without this, as we have seen, variation can lead to fragmentation.

## Tolerance Is Not Enough

In an evolving, or quantum, society, mere *tolerance* of the other is not enough. Both conflict and tolerance keep the other at arm's length. Both stress that he—the black, the Asian, the Muslim, the Jew, etc.—is *other* than myself. He is no part of my experience except insofar as he impinges upon me. He is no part of my identity, of my "I-ness." He does not get inside my skin. As sociologist Zygmunt Bauman expresses it,

> Tolerance as 'mere tolerance' is moribund. . . . It just
> would not do to rest satisfied that the other's difference does not
> confine or harm my own. . . . Survival in the world of contin-
> gency and diversity recognizes another difference as the necessary
> condition for the preservation of its own. . . . Tolerance is ego-
> centred . . . there is a lot of cruelty that tolerance, through the
> lofty unconcern it feeds, makes *easier to commit*.[7]

It is this whole concept of the "otherness" of the other, of
the arm's-length distance of the other, that a quantum view of self
and of our life together in society calls into question. It makes us
look again at the nature of identity and relationship and makes us
ask where, if at all, it remains meaningful to draw a boundary
between myself and the other, between "us" and "them." The
process of evolution, too, raises this same question. To what extent
can we, in an unfolding process, distinguish one element of that
process from another? To what extent are separation and "otherness" meaningful?

We have seen that if human consciousness is quantum mechanical in its origins, identity, the very core of the self, has both
a particle and a wave aspect. The particle aspect of my self is my
"I-ness," that part of me that is a unique and identifiable pattern
with my own peculiar characteristics, my own "voice," my own
style. It is that aspect of me that makes me a soloist in my own
right. The wave aspect of my self is my "we-ness," the part of
me that is evoked through my relationship to others and that is
literally interwoven with the being of others. My wave aspect is
my public aspect, the dance to which my solo movements contribute and through which they both take on a larger meaning and
evolve.

The culture of extreme individualism associated with the old
mechanistic physics recognizes only our particle aspect, our "I-
ness." In this culture we perceive ourselves, our group, our kind,
as separate islands of experience surrounded by clearly defined
boundaries. We perceive ourselves, to use my earlier analogy, like
Newtonian billiard balls that bump into and clash with one another
(conflict) or, at best, avoid clashing (tolerance). We are, as Freud

described it, objects to one another: "I am a self to myself, but an object to others. To others I am a thing, a 'what,' and others are objects to me."[8] We are self and other.

But on a quantum view of the self, even this I-ness, or particle aspect, is radically different. It is less opaque, less rigidly defined, less static. The boundaries of the quantum self are more entangled. I-ness is on an ever-shifting continuum with we-ness.

If we return for a moment to what might be the actual physics of the mind, we can recall that the core of our quantum-based identity is more akin to a whirlpool than to a billiard ball. It is what physicists call a self-organizing system. Such systems are shifting patterns of dynamic energy in constant creative dialogue with their surrounding environments. They take material, experience, information from their surroundings and weave these into the fabric of their being. Each of us does this from moment to moment as we live our daily lives—our minds reach out and take in the experiences that we are becoming. At our core (our particle aspect) we are a recognizable but ever-changing pattern. At our "periphery" (our wave aspect) we are a teeming web of relationships. Each of us is *both* self *and* other.

In our personal lives we recognize this dual nature of identity as a daily reality. Each of us who experiences intimacy with a loved one knows that the boundary between self and other is a tenuous thing indeed. In any intimate relationship it is often impossible to say where "I" end and "you" begin. Each of us evokes qualities in the other that neither had realized before. We are as the dancers and the dance. I believe it is the same at the level of the groups or the communities to which we belong.

Communities, too, like individuals, have a sense of identity or "group consciousness" that each recognizes as its own. Each has its own culture or set of "language games," as the philosopher Wittgenstein would have called them. These may be subtle or minor, as in the local customs and practices that typify a street gang or football club, or they may be much more pervasive, as in the traits that distinguish religious, ethnic, or national groups. But each is a recognizable pattern for structuring some or all elements of social reality. Each, to use the language of physics, is a self-organizing system in its own right. Each, I believe, functions according to the same physical dynamics (albeit at a more complex level) that form individual selves.

Communities, too, can be perceived as dual-aspect quantum systems. Like the individuals of which they are composed, each community has a "core" and a "periphery," a particle aspect and a wave aspect. Each is to some extent both self and other, the other or others with which it shares a public space.

Earlier, I described a mechanistic model of how different communities in society relate as a system of moving pendulums attached to each other by springs of varying strength. In this model, each pendulum (each community) is separate in itself and related to others only through force. A sense of the other as threat, of the other as someone who might deflect me from my course, pervades this picture. The closest that mechanism can get to any form of pluralism is a system of mutually agreed tolerance—i.e., liberal coexistence.

In looking for a good physical analogy of a pluralist quantum society, I think particularly of the very powerful image of the benzene ring. A benzene ring is very different from a system of pendulums. It is a specifically quantum structure whose carbon electron shells have become entangled. At certain levels there is no isolation, or separateness, of parts.

A benzene molecule itself, if looked at in "still life," consists of six carbon and six hydrogen atoms living as a group. In any ring we may substitute some other atoms (called the "R") for some of the hydrogen. This substitution gives us a new compound that contains a benzene ring within it, but it has the same group dynamics as the original. This group sustains its own peculiar pattern of group identity. But in fact no benzene-type molecule lives in a single state. Rather, two (or more) *possible* benzene molecular arrangements (substructures) are held together by an entangled "cloud" of six shared electrons that draw them into a characteristic ring relationship. If the substructures were not cojoined in the ring, each electron would belong to a specific place in the ring, like the thoughts belonging to specific individuals when they live alone. But in the ring, none of the shared electrons is localized (none belongs to any one carbon atom). Their position can't really be pinned down.

FIG. 8.2 THE BENZENE-TYPE RING (WITH ITS TWO SUBSTITUTE "R" ATOMS) ISN'T FIXED IN EITHER OF THE TWO POSSIBLE STRUCTURES, NOR DOES IT OSCILLATE BETWEEN THE TWO. RATHER, IT IS A QUANTUM SUPERPOSITION OF (MAINLY) THESE TWO POSSIBILITIES.

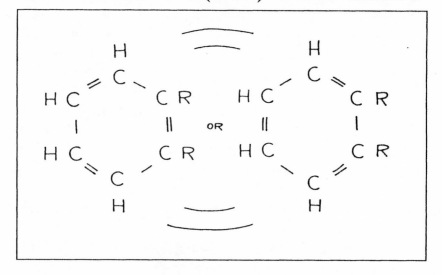

They are now like the thoughts shared by a community, or a culture.

Because of the shared electrons, which cause the identities of the constituent possible group substructures to overlap, the benzene ring itself exists in a quantum superposition of possible localized structures, just as Schrödinger's cat exists in a superposition of alive/dead states (Fig. 8.2). Physicists call this a "resonance state." It is more stable than any one molecular arrangement would be on its own. By overlapping and entwining their carbon electrons in this way, by creating a shared electron domain, the constituent (possible) molecular arrangements in the benzene ring are able to exist in a lower energy state.

In fact, there are many resonance ring structures in quantum chemistry. Some of the more complex ones have substructure groups consisting of different sorts of atoms, so that each group has its own characteristic "individuals."

Adapting the benzene ring to the language of a quantum society, we would say that the various component atoms share a "public" sphere (the shared electron domain). This sphere is not

FIG. 8.3  A COMMUNITY OF COMMUNITIES—THE BOUNDARIES OF
EACH OVERLAP THROUGH SHARED POTENTIALITIES.

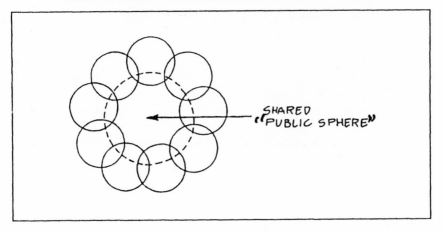

neutral, as in liberal coexistence. Rather, the public sphere is the
*shared reality* participated in by each of the ring's atoms. This shared
reality is the stuff of each individual atom's substance; it allows
each molecule to be itself. And it is a creative reality. Through
participating in it each component atom acquires characteristics
(stability) it did not have on its own.

We might think of each component atom in a benzene ring as
being like a component community in a pluralist quantum society.
The benzene ring itself with its superposition of possibilities is a
physical equivalent of the "community of communities" spoken
of by Jonathan Sacks in my opening quote (Fig. 8.3). Such a quantum model for pluralism is radically different from a mechanist
one. In the quantum model, the sharp, mechanist boundary between self and other gives way to a more fluid overlapping and
entwining of constantly shifting dynamic patterns. The mechanistic
perception of the other as threat gives way to a perception of the
other as one who evokes my own latent possibilities. The quantum
other is both (an aspect of ) myself and my opportunity—my opportunity to grow and to evolve, my opportunity to realize my
own potential self. Seeing the other as such takes us well beyond
an attitude of tolerance. *The quantum other is my necessity.*

We saw in Chapter 4 that the quantum self *is* its relationships.
It is defined in terms of its relationships—to others, to its culture,

its group, to nature. At every level that we search for an "inner self" we find further relationships (or roles—mother, wife, writer, etc.). We find the values and the systems of belief that are associated with these relationships, or with events and relationships in one's past. Most of us never discover a transcendental, or featureless, self apart from those relationships.* Each of us is like a fish swimming in a sea of inner and outer meanings, each associated with some relationship. The richer the range and variety of these relationships, the richer and more quickly evolving the self, providing that we can integrate them meaningfully.

At the most primary level of quantum reality, the necessary relation between a kind of natural pluralism and the evolution of quantum systems is obvious. When any quantum system evolves, we saw (Chapter 2) that it does so by throwing out a plethora of possible moves, or "virtual transitions." Earlier I called these "feelers toward the future." Each of these possible moves expresses one aspect of the system's potential for growth. Each is one "face" of its underlying reality, and all are necessary for the system fully to explore its options. As the system evolves, its various possibilities get into dialogue (in the language of physics, "interfere") with each other and with the surrounding environment. Each possibility is an integral part of the unfolding process that will enrich them all.

The benzene ring analogy just discussed applies to the social context if *we* are the quantum systems. We have seen repeatedly that a quantum system consists not only of measured actualities but also of shared potentialities and that these potentialities affect the dynamics of the system. The atoms in the benzene ring have the potential to relate in different ways—and they express all of these simultaneously. As individuals, we are not only what we *are* or *do,* but also what we *might be* or *might do.* If we share our wishes, fears, dreams, and images, implicitly or explicitly, we become

---

*Some mystics claim to reach this stage, but they usually describe the transcendental self as nonpersonal, as at one with some higher reality. I would argue that the transcendental self, too, is defined in terms of relationship, a primal relationship to "God," or to the ground of existence. In Chapter 10 of this book we will see that this ground of existence can be identified with the quantum vacuum.

bound together into a group that transcends any one definition of itself in actual behaviorial roles.

I believe this same pluralism drives the evolutionary process in society at large and enriches each community that participates in that process. The combined habits, mores, styles, and traditions of each community or group is one "feeler toward the future" in society's overall evolutionary drama, one expression from many possible ways of being. In religious language, we might say that each community is an expression of one of God's possibilities. If we fully understand this, it must transform our relation to the others with whom we share our social space. It must transform our attitude to that social, or public, space itself. The shared public space where diverse groups or cultures meet is the common pool of our evolutionary potential. It is, more than that, the common ground of our being. It is an integral part of each and every one of us, of each and every community, an integral part of our growth and vitality. We need that shared public space, that shared public *reality,* fully to be ourselves.

## A Celebration of Diversity

Each one of us in our personal lives recognizes the value of diversity. We know that we thrive on choice, on there being a wide range of possibilities from which we can choose. Our freedom only makes sense in the presence of such choice. We know that we often grow when exposed to new people, new ideas, or new situations. This is why we travel, taste new food, or send our children away to university. This is why many of us prefer to live in cosmopolitan cities. But the mere presence of diversity is not enough.

Diversity on its own can be like my young daughter and the pile of Christmas presents. It can shock and threaten because it has the potential to fragment or to relativize our own experience. It can lead to a cacophony of voices rather than to the "continuing conversation" that Jonathan Sacks so rightly says we need. So how can we encourage that conversation? How can we, as I put it, cultivate our shared public reality and thus turn the modern fact of diversity to our common good?

In later chapters I will discuss how, practically, we can do this through our institutions, through our politics or our city designs or in the way that we educate ourselves and our children. But here I want to concentrate on what each of us can do as individuals to cultivate a creative dialogue with others. How can I, as a Christian or a Jew, a black, a white, or a Muslim, live my personal life in a way that transcends the mechanistic boundary between self and other? How might I, as a citizen of Los Angeles or any other large urban center, reach into or beyond the ghetto?

As I ask this question, I recall two recent incidents that left a deep impression on my own thinking about conflict between people. The one was a black man from Los Angeles's burned-out south-central district who looked on in despair as President George Bush was rushed through the remains of his neighborhood surrounded by a crowd of security men. "I wanted him to listen to me," the man said, looking at the president's back. "I just wanted him to hear about my feelings."

The other incident took place at a major university in England. For several years a well-known professor had been the focus of demonstrations and threats by animal rights protesters because of his research work on small mammals. One man in particular succeeded several times in discovering the professor's often-changed unlisted home telephone number. He used this information to phone frequently, often in the middle of the night, shouting threats and obscenities before banging down the phone. During one of these calls, the professor had a chance to break in. He suggested to the man that they meet and the man agreed. At first the meeting was like the phone calls. The man was upset and aggressive. But the professor listened. He said later, "I found myself feeling that I understood him. I understood why he was so upset and how he felt."

This second incident became a conversation. The protester did not change his point of view or his objections to the professor's experiments, nor did the professor give up those experiments. But the phone calls in the middle of the night stopped.

There is an old American Indian proverb that says, "Great Spirit . . . grant that I may not criticize my neighbor until I have walked a mile in his moccasins." Simply listening to my neighbor is the most powerful way that I can do this. Listening, opening myself to his feelings and his point of view, accepting that he *does* feel the way that he feels, dissolves the boundary between us. Listening allows the other to get inside me, to become a part of me. It allows us to "meet." Photons in the quantum laboratory do this when they are "introduced" in some "social" setting (brought together in a material medium). Their wave fronts become entangled and they share some aspects of their identity. We, too, become entangled, partners in a creative public space, when we meet. I recognize that the other, too, is an unlived potentiality deep within myself.

There are many forms that creative listening might take. It may mean, as in the case of the animal rights protester and the professor, literally listening. This was all the black man in Los Angeles asked of his president. It may mean opening oneself to the other's way of life—reading his poems, listening to his music, attending his ceremonies, eating his bread, "wearing his moccasins." Among the head-hunting Kiwi-Papuans of New Guinea it takes a very extreme form.

When one village of Kiwi-Papuans wants to end a conflict with a neighboring village, it lays a branch across the path leading to the village of the former enemy. If the branch is taken up, the men of the first village approach, taking their wives with them as a sign of good intention. There is a meeting, gifts are exchanged, and the men of the two villages break each other's beheading knives. At night, the hosts sleep with the visitors' wives in order to "put out the fire" of conflict. The next day, the same ritual is repeated in the other village, and peace is declared.[9]

This Kiwi-Papuan ritual seems a peculiarly "quantum" form of conflict resolution, though I doubt many Western women would be prepared to reenact it. I have included it because it amuses me, and because it is a very graphic example of letting the other get inside one's self, of letting the other become flesh of one's own flesh. I believe the quantum nature of our consciousness, the fact that each of us has a wave aspect and a particle aspect, makes this commingling possible without its more literal Kiwi-Papuan interpretation. An appreciation of the other's creative role in the formation of my own identity and in my own evolutionary process makes it necessary if I want fully to become myself.

Ironically, if the physical basis of consciousness is a quantum self-organizing system in the brain, we can see that "listening" to the other, opening ourselves to him and taking him inside our boundaries, is both the most *necessary* and the most *difficult* task of mind. I need the other fully to become myself, but at the same time something in me often resists dialogue with him. Any good therapist or counselor or negotiator knows that simply getting two sides of any quarrel (or of any individual self) to listen to one another is a very difficult thing indeed. Habits of thought, preformed opinions, and stereotypes of the other often get in the way. The physics of self-organizing systems makes it clear why this might be so.

Self-organizing systems, I have said, are like whirlpools. They take material or information from the surrounding environment and form it into a dynamic pattern. In the case of biological systems, they take material and form it into patterns of tissue or organism. In the case of mind, information is formed into patterns of thought, patterns of meaning. Our minds take information and experience in and literally weave it into a *picture* of our world or a picture of other people. Forming such pictures, making sense of experience, is the task of consciousness. While we are forming the picture, we are learning—about our environment or about the other—and we are *open*.

While we are forming the pattern, or narrative, of thought, we are almost greedy for experience or information. We need it to complete our world picture. Childhood is a constant process of such learning because the child needs a great deal of information and experience to form the complex picture that is the adult's worldview. Thus we often describe children as "open to experience" or "fresh." They are less likely to become locked into habits of thought or into stereotypes of the other. They are still struggling to form a narrative.

But once we have reached adulthood and have formed a viable picture of the world, we are less driven to take in new information. Simply maintaining the pattern of existing thought requires less energy than constantly having to weave new information into the pattern. Habit and stereotype become tempting because they are easier. A picture of the other that fits into my existing categories of thought, into my existing narrative, requires less work—literally less physical energy. Thus adults can often be less open to new experience or learning. Adults find it more difficult to listen to others. We become locked into a self-perpetuating pattern of thought and meaning. Challenge to this pattern can feel threatening or even painful. It is a challenge to the whole fabric of self and world that one has woven.

But self-organizing systems *thrive* on challenge. As I described them earlier, such systems are poised on the brink between order and disorder. Too much disorder, and the system is torn to pieces. But too much order and it "dissipates." A self-organizing system that "relaxes," that falls out of dialogue with its environment, runs down. A mind that ceases to take in new information grows stale, less alert, less conscious. Habit and stereotype are not just a threat to my attitude toward the other. They are a threat to *me*, to the vitality of my own mind. We see this often with people who are subjected to routine work. The same threat exists when we subject ourselves to routine social relationships.

Thus listening to the other, making an effort to go out and meet the other, is both difficult and necessary to the kind of physical systems that we are. It is tempting to live within the world picture or narrative of our own group, our own kind, and difficult to extend that narrative. But the vitality and evolutionary future of both ourselves as individuals and the groups to which we belong depends upon our making the effort.

In fact, many natural (physical) systems spontaneously organize themselves to a state on the borderline between order and chaos. This is the state that physicist Ilya Prigogine describes as "organized criticality." A simple example is a pile of sand, onto which more sand slowly falls at the top. The pile spontaneously assumes a conical shape, at a critical slope where the addition of even one more grain may cause a small or large avalanche. Such systems are very sensitive to new stimuli. Some features of brain dynamics seem to be like this.[10] To "live at the edge" like this is to be constantly fresh, though potentially unstable. This is the rule for children, creative artists, and some borderline psychotics, all of whom have a greater than usual capacity to see life anew and to respond spontaneously, though often at the risk of being like the pile of sand. For the majority, life lived *constantly* at the edge may be too demanding. Some habits are necessary, too, but reliance on them is easily overdone. If we are to live full, "quantum" lives, we must always balance habit with an openness to new experience.

In many mystical traditions, there is the belief that some form of temptation is necessary to redemption. (In the language of physics, temptation is a potential instability.) The Jewish mystics hold that the man who lives his entire life couched within the safe confines of his own observant community is less likely to experience true holiness than the man who has wandered among foreign lands and sampled foreign ways. This is the basis for the parable of the Prodigal Son. It is also a spiritual way of expressing what I have said about the structure of consciousness and our relation to others who are different from ourselves. For consciousness to be "redeemed," for our selves and our communities to be made whole, we sometimes must reach out to "foreign lands" and "foreign ways" and take what we learn there home to our own communities. The fruit of such expeditions both enriches our experience and gives us the necessary elements for a dialogue with the other.

Knowledge of the other, experience and meeting with the other, *listening* to the other, and learning his "language" are all necessary ingredients of dialogue in a pluralist society. All are necessary to turning diversity to our mutual evolutionary advantage. But dialogue requires something else, too. Dialogue requires that each of the parties in a conversation has a recognizable voice, something of his or her own to contribute. As Paul Ricoeur says, "In order to confront a self other than one's own self, one must first have a self."[11] In quantum language, we would remind ourselves that each person or each community has both a "wave aspect" and a "particle aspect," and that both are equally primary.

If we emphasize only the particle aspect, the private identity that each of us experiences as members of our own community, pluralism is in danger of becoming a Tower of Babel, a cacophony of voices each expressing itself in a language that others cannot understand. As we have seen, such cacophony does not give us a public conversation, it does not give us a shared public reality. But if we emphasize only the wave aspect, the common language that each of us can understand, pluralism is in equal danger of becoming a vague syncretism, a vapid "stew," the bland ingredients of which excite no one's palate. The attempt to invent Esperanto* suffered this fate. It was a false public language with its roots in no culture.

In stressing *either* the particle aspect *or* the wave aspect of our different communities, we have what Jonathan Sacks points out are "two incompatible views of a plural culture. One sees it as a place where many traditions meet and merge. The other sees it as an environment where distinct traditions can guard their separate integrity."[12] The one gives us the "melting pot," or a vague syncretism, the other the Tower of Babel. Both are outmoded, mechanistic alternatives.

In a quantum society, we would stress *both* the particle *and* the wave aspect, stress that these are an inseparable duality. In our individual lives we recognize that we are both private selves and public persons, and that there is no hard and fast boundary between the one and the other. In our communal lives we would stress both the individual integrity, the customs, the traditions, and the language of our own community, *and* the public sphere which is our common further reality—the shared customs, shared traditions, and shared language of our common culture. Ultimately, if nothing else, we share our common humanity.

While reflecting on how we can balance a loyalty to the integrity of our own culture, our own faith or tradition, with a recognition that it is but one voice in a wider public sphere to which we also feel loyalty, I think of the secular example of a community of physicists. There are many *branches* of physics, each with its own specialized knowledge and vocabulary and sometimes with its own larger vision of how the physical world works. Often physicists in one specialized field can't understand the language of those in other fields. Sometimes it is even the case that their theories and experimental results contradict one another. Each physicist is proud to be a member of his field and to hold the individual expertise that it confers, but at the same time all physicists are

*A syncretic official language intended to facilitate international communication.

conscious of being members of a wider community of scholars who share the same basic interests and adhere to the same set of ultimate values (a loyalty, for instance, to scientific method). All hope that one day the insights from each special field will be integrated into a broader, more unified picture of physical reality.

Thus we don't fall into cultural relativism through our encounter with other cultures. We remain true to ourselves, committed to our way, our customs, our values, while entering into a creative dialogue with others. We will change through this, as a matter of evolutionary course, but that change, too, will be part of our evolving history.

To live fully both aspects of our dual social reality we must, to use Jonathan Sacks's expression, become "bi-lingual." We must develop what he calls a facility for "first and second languages." The one is the ongoing, developing, but always recognizable language of our group, the other is the language with which we carry on our public dialogue, a conversation expressing our many points of view. The second language, the public language, will evolve as our dialogue continues. It will be rooted in our many cultures and yet have a life of its own, "a community that embraces many communities," or a dance through which many dancers express their own unique styles.

# 9. The Fifth Son: Society's "Invisible" People

*There are four sons who take part in the Passover ceremony.*
*They ask The Four Questions. But there is also a fifth son. The*
*fifth son is the son who is not at the table. Without him, the*
*ceremony is incomplete.*

Jewish Parable

Two years ago, my husband and I were having lunch in an outdoor café located in a churchyard alongside one of Oxford's busiest streets. We were reveling in the brilliant sunshine, good food, and a heady conversation about religious insight.

Suddenly a woman grabbed a free chair at our table and sat down. She sat down belligerently, glaring at us and our food. She was both drunk and mentally unstable. We asked her to go away. She tore at a piece of stale bread, eyed us, and spat out, "Won't go 'way!" As I went to get the manager, the woman grabbed for my husband's plate of food. When he tried to stop her, she swept the plate from the table. It smashed against the ground.

While the manager tried to ease the woman away, a young man approached the table. He, too, was drunk and mentally unstable. He reached toward me and tried to grab my piece of cake. As I gestured to stop him, he began to shout. Between a stream of hate-filled obscenities, he cried out, "Good, be afraid, you *****. I like to see you posh people upset."

The woman leered. She had a companion now and was encouraged. While the manager did his useless best to calm the situation, she grabbed my cup of coffee and threw it in my lap. The man smashed my plate with his fist.

My husband and I were advised to withdraw inside the café. We were given new food, a profuse apology, and showered with sympathy by other customers who had witnessed the scene. I did, of course, feel very upset, and defeated—defeated by an unaccustomed powerlessness to defend myself against the violence. I also, when I had calmed down, felt deep shame. I was shamed not by my clear right to enjoy lunch in a garden café, but by my knowledge that had those two people not assaulted me and thrown my food to the ground, I would never have noticed them. If the woman had gone away when we first asked, I would not have given her another thought.

That scene in the café has stayed with me. I remember it each time I pass a beggar's outstretched hand on Oxford's central streets. I remembered it when I saw the huddled and broken figures that line nearly every street in downtown San Francisco. I thought of it when I drove past the miles of tenement blocks en route to John F. Kennedy International Airport or those same bleak structures that encircle Paris and Madrid. I thought of it when gangs of violent young people, rootless and unemployed, smashed their way through London's West End, and again when Los Angeles burned. I feel it every time I avoid eye contact with a potentially demanding stranger who approaches me on the street.

In some of these cases shame, and a good dose of liberal guilt, may be appropriate. Perhaps I could do more, or give more. Perhaps it is wrong or unjust or unfair that I should have so much and live so well when there are these others without. But now I realize, too, that something other than shame is the source of my unease. Something deeper, more personal, more painful to *me*, is in question when I see these others who have "no place at the table." In some very direct way, I know that *I* am threatened by their absence, that they are in fact some missing or overlooked part of *me*, and that somehow, as in the parable of the Fifth Son, their absence leaves my own ceremony incomplete. Their absence leaves my *self* incomplete.

There are various familiar ways we describe those who live at the bottom of society, or "beyond" society. Marx called them the Lumpenproletariat to distinguish them from the oppressed working class. The more favored modern term is the "underclass." Americans, less kindly, often dismiss them as "scum," denying not only their place in society but perhaps even their humanity. They are not all the same. Some, like those who disturbed my lunch, are mentally unstable. They can't easily cope with the complex demands of an independent life in impersonal large cities. Others are children born out of wedlock and raised, haphazardly, without the benefit of family. Some are school dropouts, illiterate, unemployed and unemployable. Others are unsupported teenage mothers, drug addicts, or criminals. Some are all these things. Most congregate in depressed areas of our inner cities and many are the all too natural products of ugly, overcrowded housing projects where welfare and crime are the bedrock of the local economy.

It is difficult to categorize all those who are among our society's deprived outsiders. Different "causes" or explanations apply to different groups, and each has its own characteristic patterns of behavior or manner of living. But something in common distinguishes most from others who are simply poor or who, by choice, live outside society's accustomed structures and mores. There is a segment of the population who share a poverty of language, a poverty of culture, an impoverished set of standards and mores rather than just standards and mores that are simply different from our own. They seem, for whatever reason, not fully to be included within civilization's pale. They form no political constituency,[1] they have few champions, and they play no positive role in the daily life of wider society.

I think, for example, of the very vivid portrayal of a Los Angeles black ghetto in the film *Boyz N the Hood*. The film's characters were not especially poor in material terms, but most lacked any moral or cultural framework for their lives. They were not immigrants who clung to some "primitive" African culture, but citizens of nowhere with no culture other than one of mindless sex and violence. The only words available to them for expressing whatever they felt or observed were "shit" and "fuck." They reminded me of many young children I overhear in British parks or playgrounds whose vocabulary is limited in the same way and who seem lost and disoriented despite their aggressive bravado.

Deprived outsiders are the other who seems wholly other. We don't easily identify with or understand them. We don't feel that we have anything to learn from them or anything to gain from knowing them. They seem, if anything, evolution's failures, not some necessary part of a process in which we are all involved. We don't know what to "do" about them. Left to ourselves we would rather not notice them. They are society's invisible people. To the extent that we do see them, they make us uncomfortable. Their very existence in such large numbers tells us that there is something wrong, some failure in the organization of our society, perhaps some failure in ourselves. We wish they would go away or change, become more like us.

We don't usually notice these chronic outsiders unless we are made to, unless their behavior directly impinges on us. President Bush had not paid a single visit to an American inner city area until the Los Angeles riots made it impossible for him not to do so. Sometimes they do make it impossible for us not to notice them. Sometimes they attack us or our property. They make it unsafe to walk down certain streets, unsafe to be alone at night, they sell our children drugs, assault us on the subway or in our cars when we stop at traffic lights, rampage through our cities. At these times, they raise the specter of violence and threat. Encountered only in this way they are, par excellence, the other as threat.

We have already seen in our discussion of meeting other cultures and other peoples that a perception of the other as wholly other, or of the other as threat, invites various mechanistic reactions in response. With other cultures, these possible reactions are conquest and domination, apartheid or segregation, or liberal co-existence. Each is an attempt to prevent or limit the other's impact on me, an attempt to defend myself against him—the paradigm of defense analysis. These same mechanistic reactions dominate to an even greater extent our essentially defensive dealings with society's most deprived. Each has its familiar counterpart in the social or political policies usually adopted and in the behavior and attitudes of each of us as individuals.

Used in response to those of our own among us who seem to threaten our way of life, conquest and domination becomes a policy of control or suppression. We put more policemen on the streets, give those policemen more powers, form vigilante groups, call out the army, impose curfews, and give tougher prison sentences. We emphasize that there is hierarchy in society (hence the unfortunate term "underclass") and use authority as an instrument to "keep down" or "keep the lid on" potentially dangerous behavior.

Where control or suppression seems too costly or too difficult or simply ineffective, we sometimes adopt the domestic counterpart of apartheid or segregation. We tacitly acknowledge "no go" areas. Certain parts of the inner city, housing projects or whole neighborhoods, are simply struck off our map. We accept that they are beyond control and we stay away. We leave "them" to get on with it, so long as they don't bother "us."

And finally, there is the welfare approach. We send in our money and our bureaucrats. We accept that some people can't or won't work, accept that teenage girls will continue to have babies out of wedlock and with no support from their men, accept that some children will refuse to attend school or can get nothing out of school, that others will take drugs. We adopt an attitude very like that of liberal coexistence. We accept the right, or the presumed necessity, of the other to be different (in this case to live outside society's usual structures) and allow him to get on with it, in his own place in his own way, with our financial support.

There may be many good reasons for giving welfare—for putting a coin in the beggar's hand or paying the unwed mother's rent. It may be there is no other way to prevent some people from starving or being homeless. It may be that some form of emergency financial aid will always be a necessary part of our response, whatever our attitude to those who have much less than ourselves. But if given with the attitude that its recipients are wholly other than ourselves, an "underclass" that should be sustained, apart from us, in ways of living that we could not stand to adopt, welfare is just another mechanistic response. It treats the other as object—the object of our charity or our pity.

We have seen, time and again, in this book that our attitudes and our perceptions are themselves creative acts. The way that we see another person or a situation can become a "self-fulfilling prophecy." We know this to be true from our daily commonsense experience (the "power of positive thinking," "a defeatist attitude"), and psychologists and sociologists often build it into their theories about self and society. An understanding of ourselves and our relationships in quantum terms makes it clear why this must be so. In dealing with quantum reality, we know that we see what we expect to see. Our perceptions or attitudes (our measurements) evoke one potentiality from the underlying reality's many-possibilitied domain. I think it very likely this dynamic is at work in our relations with those who live on the fringes of society.

I think that if our attitudes (and the consequent policies that we adopt) toward society's most deprived outsiders are mechanistic attitudes, if we perceive them as wholly other, as *objects* of our pity and fear, as a problem to be overcome or a threat to be defended against, these attitudes may evoke the very reality they conjure up. In particular, they may evoke the behavior that we most fear, the rage and the violence that do indeed threaten us, our property, and our daily activities.

Most of us, in ourselves, are complex communities of subselves or subpersonalities, pockets of awareness and experience and reaction, each with its own diverse needs and degrees of maturity. Some of these subselves are less well integrated than others, some may be "dark corners" of the self, ineffective or injured, split-off aspects of our overall personalities. In Freudian psychology, a whole range of our emotions and instincts is perceived as dark and tempestuous, "animallike" or uncivilized. These are the forces of the Id, the basement of the self, our own personal "underclass." And yet they are, undeniably, parts of ourselves.

When I write about these lost or unwanted parts of the self that each of us harbors, I am reminded of the powerful image conveyed by the Vietnamese poet Thich Nhat Hanh in his poem "Please Call Me by My True Names." Feeling extreme pain at

the suffering of some and yet able to see himself as one who might
cause such suffering, he writes,

> I am the mayfly metamorphosing on the
>    surface of the river,
> and I am the bird which, when spring comes,
>    arrives in time to eat the mayfly.
>
> I am the frog swimming happily in the
>    clear water of a pond,
> and I am also the grass-snake who,
>    approaching in silence,
>    feeds itself on the frog.
>
> I am the child in Uganda, all skin and bones,
>    my legs as thin as bamboo sticks,
> and I am the arms merchant, selling deadly
>    weapons to Uganda.
>
> I am the 12-year-old girl, refugee
>    on a small boat,
> who throws herself into the ocean after
>    being raped by a sea pirate,
> and I am the pirate, my heart not yet capable
>    of seeing and loving.

In each of us, Hanh reflects, there is the potential to be all these
things, to be all others. If we realize that we might have become
the pirate, we will be less hasty to judge. If we shoot the pirate,
as well we must, we nonetheless shoot something in ourselves.
"Please call me by my true names," he concludes,

> so I can hear all my cries and all my laughs
>    at once,
> so I can see that my joy and pain are one.
>
> Please call me by my true names,
>    so I can wake up,

and so the door of my heart can be left open,
the door of compassion.[2]

Freud himself, as I mentioned earlier, was unsure what to do
about the unwanted parts of the self consigned to the Id. If we
repressed them—failed to recognize them, denied their existence
or their force—he thought, they had the power to make us ill
(neurotic). If, on the other hand, we let all these dark forces run
rampant, they would, he feared, destroy civilization. His partial
answer was a mechanistic one. We should recognize the Id but
fiercely control it. The Ego, the socially acceptable and rational
part of the self, should act as a firm policeman to rein in the Id's
destructive powers. We recognize in Freud's dilemma our own
mechanistic notion that we must suppress and control society's
"underclass" lest it, too, run rampant and destroy us.

In dealing with the "dark" corners of ourselves—our childish
needs, our infantile rages, our undigested pockets of hurt and pain,
our unseemly jealousies or unwanted sexual fantasies—many of
us respond as we do to others among us who seem socially un-
acceptable. Our first reaction is often denial. We would rather
not notice these aspects of ourselves. We treat them as invisible
parts of ourselves or no part of ourselves at all. Our shame that
they are there or our fear that they might corrupt us or take us
over leads us to push them away. But as Freud noted, this denial
can bring about the very destruction it fears.

A part of the self that is denied, that is cut off from recognition
and dialogue, festers. It can become a pocket of trapped psychic
energy that makes us ill or a kind of personal time bomb that
threatens to break out unexpectedly with violence. We see this
same dynamic at work in our personal relations with others. If
someone is angry with me and I deny that his anger has any jus-
tification, that anger may reach a point of violent outburst. If, to
punish one of my children, I ignore him or her, the child becomes
violently agitated. Children cannot bear to be treated as "invisible." As Robert Bellah and his team emphasize in *The Good Society,*
"Infants who do not get attention, in the sense of psychic interaction and love, simply cannot survive, even if they are fed and
clothed."[3] Bellah cites a recent psychological study that goes fur-

ther in showing that children who get a high level of attention from their parents do much better in life, are more cooperative, friendly, and socially well adjusted than others whose parents paid them or each other less attention.[4]

Each of us can probably recall some instance of feeling a violent impulse in response to being ignored or treated as a "nonperson" by a teacher, a boss, or a bureaucrat. Much of the venom that inspires modern feminism follows from some women's feeling that they were perceived merely as objects, of men's desires and manipulations. How much more potent must be this potential for maladjustment and violence in people who are ignored or seen only as objects across the board, people who are disregarded as being no part of "us."

A quantum understanding of the physical dynamics of consciousness can help us to understand why violence is a natural response to being cut off or ignored. If consciousness is an open, self-organizing system, it literally depends for its continued existence upon a constant input of relationship. In quantum terms, we *are* our relationships. We are relational wholes, some of whose qualities only come into existence when our being overlaps with that of others, like the quantum dancers we encountered in Chapter 5. Denied that overlap, that place in the dance, we are deprived of the very substance of our conscious being. Such deprivation is real, physical, and would be physically painful, just as a deprivation of food is painful. Such pain can evoke a violent response.

Within the brain itself, this physical dependence upon relationship is mirrored in the coherent neuron oscillations that give rise to consciousness. Only those neurons that oscillate in synchrony with their neighbors—that overlap and combine their identities—get caught up in the emergent reality of consciousness. Neuron modules isolated by illness, damage, or underuse have no part in this process. On the contrary, they may interfere with it, dampen it down, and leave it less whole.

Socially, our most deprived outsiders are like unused or damaged neuron modules in the brain. They are not parts of some larger, synchronous whole. Unlike minority groups with clearly defined identities and mores, they have nothing that draws them together as a group. They aren't organized (nor, I believe, could they be peaceably) and they have no leaders who speak for them. This is why they form no political constituency. Violence is the only catalyst through which they ever function coherently.

Through violence, as members of a mob, these chronic outsiders who live most of their lives in isolation suddenly acquire a higher, group reality. They acquire the further identity, the sense of belonging to something larger than themselves, the potent elixir of relationship that most of us experience at least somewhat in the normal course of our insiders' lives. Such belonging makes them "feel good," as the football hooligan I quoted earlier put it. "You become part of the crowd and it gives you something." Thus violence can draw them, to the football stadiums or to the inner city streets. It can be the only means available for assuaging the pain of being wholly other or wholly outside.

On a larger scale, this same mob psychology underlies the rise of narrowing and intolerant nationalism within our increasingly pluralist democratic societies. One British political scientist has called it the "scavenger" of democracy.[5] When a majority or dominant group feels its hegemony threatened, it often seeks to redefine national identity in terms of its own limited perspective. The "real Americans" or the "good Germans" are those who are like "us," those who wear "our" kind of clothes or who hold our ideas or values. Once again a potent sense of belonging to the group can all too often express itself through violence, or through "a bovine will to simplify things."[6] The nationalist cannot find his own identity in the larger, more complex web of heterogeneous society. He sees all others as threats.

There are two different but complementary reasons, then, why the physics of consciousness would lead us to expect a violent reaction from those who are excluded from society. The one is the physical pain of the exclusion itself and its accompanying rage. The other is the allure of a sense of belonging available only through mob activity. Both, I believe, are exacerbated if not wholly evoked by our own mechanistic attitudes toward deprived outsiders and by the institutional responses to which those attitudes lead.

## *There Are No "Others"*

From earlier discussion in this book about the nature of quantum identity and the physics of a quantum society, it is obvious what I might say about the perception of any other as wholly other. Quantum reality itself is a nexus of unbroken wholeness. Every system that makes up that reality is entangled with every other, it derives essential parts of its identity from that entanglement. In the same way, quantum persons have no hard and fast boundaries, no essential atomistic or "transcendental" cores that can be defined apart from our relationships. The deeper and more extensive our relationships, the richer and more resonant our persons.

On a quantum view, each potential relationship that I encounter is a potential part of myself. Each "other" is a potential part of myself, someone without whom I am less than I might be. This is easier to see and accept when that other is someone with his own valuable culture and traditions, a member of some other community or ethnic group. But I think it is also true of those with whom we find it more difficult to empathize. The whole thrust of humanistic psychology (Carl Rogers, R. D. Laing, Thomas Szasz), of course, leads to the view that it is ultimately possible (and desirable) to empathize with anything human. This acceptance of damaged parts of the self or of damaged others is in sharp contrast to medical psychiatry's bias that they are diseased. The humanistic psychologist meets the damage with empathy, the medical psychiatrist with "treatment."

In the brain it is clear that the whole brain works better if damaged or unused neurons can be caught up in the synchronous neuron oscillations that give rise to consciousness. The same is true at the level of the self. My whole self is diminished if I isolate or ignore the "dark," injured, or immature aspects of myself. I will be more healthy, more "whole" as a person if I can recognize and integrate them. Society, too, the public sphere or "dance" in which we all participate and through which we derive so much of our identity, is diminished by the absence of those who are excluded. Just as the body is subject to illness from excluded parts of the self, society is made sick, is subject to violence, from those whose pain causes them to lash out toward that which excludes them.*

Understood in terms of our quantum nature, then, there is a mutual need between myself and the deprived other and between society and its excluded members. The outsider needs me to recognize him, to include him. He needs society to draw him in and make him a member of the dance. But I need the outsider in order to be fully myself. Society needs him for itself to be healthy and whole. There is between me and these deprived outsiders no relation of otherness. They are parts of myself.

The moral shift required by a quantum relation to deprived outsiders is a radical one. If I view the other mechanistically, as an other, the closest moral bond that I can have with him is an attitude that I am my brother's keeper. This is compatible with the best impulses of modern liberal morality. It is compatible with a welfare approach to those who are deprived. But in a quantum society I am more than just my brother's keeper: *I am my brother.*[†] He is flesh of my flesh and stuff of my same substance. If my brother is in pain, that pain is mine. If he is hungry, I am hungry. If he lives in exile, then something of me, too, is absent from home.

If I adopt the quantum attitude that the deprived outsider is an integral part of myself and my community, then my response to him will be different. I will want to *treat* him as a part of myself, include him as I would want to be included, nurture him as I would want myself or one of my children to be nurtured. But immediately there seem an overwhelming array of practical problems.

To begin with, the scale of the problem is enormous. In every major city there are thousands of people living rough and untold thousands more living in conditions of considerable social deprivation. I can't, unless I am very wealthy indeed, put a dollar in every beggar's hand. I can't myself house the homeless or care for all the neglected children. It would be naive in the extreme to think that I could still a rampaging mob with an expression of all-

---

*Some neurons in the brain, of course, may be damaged beyond repair or integration. The same may be true of a few damaged individuals in society.

[†]There is a more lengthy discussion of quantum morality in *The Quantum Self,* Chapter 11.

embracing love, or even present myself defenseless and with smiling acceptance toward a single would-be mugger. These are large-scale social problems that we feel must be met with an institutional response. In the absence of that response, it seems, even the most profound transformation of my personal attitudes toward deprived outsiders leaves me with "the difficulty of being a good person in the absence of a good society."[7]

But here it is important to recall our earlier discussion of the dynamic and mutually creative relation between individuals and institutions (between dancers and the dance) in a quantum society. Institutions are, in Durkheim's words, our "collective patterns of thinking, feeling and action." Others describe them as "socially shared and enforced norms" or "patterned ways of living together." However we describe our institutions, they are not separate from us. They have a reality of their own, but that reality emerges from our collective activity as individuals. There is no hard and fast boundary that places the individual on one side and the institution on the other.

In our quantum brain model of society (Chapter 4), institutions were like the networks of memory laid down in response to the brain's coherent neuron oscillations. They were formed by those coherent oscillations, by consciousness, but in turn feed back to act upon future oscillations. The dialogue shapes both memory and consciousness. In society, we saw that this is like the dialogue between individuals acting together in ad hoc groups (the social equivalent of the temporary synchronies between neuron bundles) and the institutions (memories) to which the activities of ad hoc groups give rise. In the words of the Bellah team in *The Good Society,*

We are not self-created atoms manipulating or being manipulated by objective institutions. We form institutions and they form us every time we engage in a conversation that matters, and certainly every time we act as parent or child, student or teacher, citizen or official, in each case calling on models and metaphors for the rightness or wrongness of action.[8]

What we regard as the rightness or wrongness of action follows from our basic attitudes and values. This is true whether the action to be taken is individual or institutional.

Thus, on a quantum view (and compatible with Bellah's way of thinking), a transformation of my personal attitudes toward society's "fifth son" is a necessary first step toward a larger-scale institutional transformation. It is, in fact, the sine qua non of institutional transformation. As Bellah says, "Our institutional problems are deeply rooted in central values, so it follows that we believe serious institutional reform in the absence of change in the central values is not likely to succeed."[9] In the end, a change in society comes down to a change in me.

As I embed my values in thought, feeling, and action, my inner transformation can resonate (get into nonlocal correlation) with similar transformations in others. More potent still, my own thoughts, feelings, and actions can *evoke* those transformations in others. As I create myself in dialogue with others, so I help to create the world I share with them. Each of us, as Jung said, is responsible for the age.[10]

So if I recognize that I as an individual need the deprived outsider fully to be myself, and that my society needs him in order to be whole, how can I embed that recognition in some meaningful action? What, ultimately, apart from having the right feelings and attitudes, can I *do* in relation to these others who are me? I think that each of us, almost daily even in our private lives, probably has some small opportunity to do something. If we are members of larger institutions—schools, hospitals, churches, companies, the police force—we can do more. At that group or institutional level, our attitudes have the chance of becoming policies.

In psychotherapy, it is known that the most important function of the therapist is simply that he *listens*. A good therapist listens not just with his full attention, but with his or her whole being. This listening has the effect of recognition. The patient doing the talking has the feeling, perhaps uniquely in his life, that someone recognizes his existence and his feelings. That recognition alone may be the most valuable part of the therapy, and it builds a strong bond between therapist and patient.

I remember, for example, a patient of my husband's whom I

once interviewed as a journalist. The girl was a heroin addict who lived as a squatter and earned what living she had through prostitution. She had no friends, and no contact with her family. Each week, she told me, on her way to visit my husband she would stop off in a major department store, steal something, and then return it for money to pay my husband. She did this despite knowing that he would treat her without charge. "He's the only one who ever listens to me," she said. "When I come to see him, it's the only time I ever feel clean." The theft, she said, was her way of "staying straight" with him. My husband, when I told him, was both moved and appalled.

Most of us are not therapists, but we do have the ability to listen, or at least to notice and recognize, others whom we meet. Perhaps recognition is all we can give—just a glance, a small smile, a meeting of eyes. It is even possible, we know, to say "no" in a way that still recognizes another's existence and need. We do this with our own children all the time. But any recognition takes us a long way from treating the other as invisible, from denying his existence or our relationship with him. Many of us are afraid to give even that, afraid that it will lead to some commitment we don't want to make or to some consequences that we fear. Some feel themselves able on occasion to give more. I think particularly of two exchanges I witnessed recently, both of which have left a lasting impression.

There is an alcoholic tramp who is always begging for coins in central Oxford. He often sings, and tells jokes to the empty air. He seems to want conversation as much as money, but most people avoid him. Last time I saw him, he was sitting at a street-side pub with a group of young men from the university. They hadn't given him a coin, they had invited him to join them for a drink. The tramp was glowing with happiness.

The other exchange happened as part of my extended family life. Some of my relatives were holding an informal open house in another city for a cousin visiting briefly from Bangkok. Most of the guests were the sort I would have expected to find at such an occasion. But then a young woman arrived with a child. She had no education to speak of, was not very articulate, and seemed socially awkward. She was, it turned out, a local woman living in a welfare hotel, whom my cousin had met while shopping. Her three older children were in foster care because she couldn't look after them. Each was by a different father, to none of whom had the woman been married.

No fuss was made over this new guest. There was no self-conscious, patronizing attention. The cousin in whose home the open house was held introduced her as she had everyone else and invited her to join us at the big table where we were talking and eating fruit. She asked after her day and took an interest in the woman's reply. The woman became fully part of the day's social activities, and she was not the only one to gain. Her presence added something for us all. Her joy at being included, and the simple yet felt experiences from her own daily routine that she shared with us, *contributed* something to the day that would not have been there otherwise. It can be difficult to express in words what we get from meeting another, but the woman's presence lent the whole day a kind of grace.

Equally, I have never been able to forget the patient of my husband whom I mentioned a moment ago, the woman who took heroin and worked as a prostitute. She is dead now. She died from an infection of the brain that was probably an early symptom of AIDS. But the time I spent with her, and my own exposure to the fierce integrity with which she insisted on dealing with my husband, has remained with me for years. Perhaps some of what touched me was the familiar "there but for the grace of God go I" sort of empathy, but I think there was more. She actually became a small part of my self, of my valued experience, that I have carried with me ever since. So too have several other of my husband's "down and out" drug-addict patients, young men who used to crouch under our dining room table making funny faces to entertain our infant children while they waited for their appointments. My husband tells me that many of his patients have this effect on him, too, that through knowing them *he* has gained a great deal.

These are just images. We are not always willing to have a drink with a local tramp or to share our home and family life with someone who has no home and whose own family affairs are in such disarray. To many the drug addict or the prostitute speak a language or hold a set of values they simply can't understand. In past, more rural times, when we lived in smaller communities, the community as a whole readily absorbed (and, where necessary, looked after) the one or two village people who had mental or social problems. In today's larger and more impersonal cities, more onus falls on the individual if there is to be a relation with deprived outsiders. Sometimes images like those I have related need to become actions.

Clearly, if our relation is to extend to many such people, we must embed it in an institutional response—through our churches, schools, neighborhood functions, employment policies, leisure facilities, police force policies and attitudes, etc. But to be effective this institutional response must include *us*. It must begin with us as individuals and with our own more "quantum" attitude toward those who don't easily belong.

I can think of one clear case where such an attitude, taken to public level, led to an effective institutional policy. For the past few years in England we have had a problem with "joy riders," teenagers who steal high-performance cars to race them on the roads late at night. Most of the youngsters who do this are school dropouts from broken homes. The usual public response is some combination of outrage and fear (people are killed by these teenage drivers), backed up by a vain police attempt to catch the culprits. But one city council, in cooperation with local car repair garages, met the problem differently. They decided to institutionalize the joy riders' activities. Parts were donated with which the youngsters could build their own racing cars, and a city track was provided where they could race them. Destructive outsiders became "members of the system" in the best possible sense. Related crime figures diminished significantly.*

It would be naive to think that any single gesture of mine, or even such gestures become policies at institutional level, could solve all the problems surrounding the pain and potential violence of deprived outsiders. Whatever we do or feel, there will always be some people who don't fit in, and some of these will always engage in crime or violence. Sometimes the only "solution" is

*The "Wheels Project," Banger Racing Track, Bordsley Green, East Birmingham.

ostracism or suppression. But both the physics of consciousness and the physics of society, if these are quantum, would lead us to expect that a transformation of our basic attitude toward others and their supposed otherness might add to the quality of both their lives and ours. If we get the attitude right, the policies and the actions will follow. Part of this more quantum attitude is linked to evolution, and to a realization that even evolution's failures (99 percent of all species are now extinct) are an integral and necessary part of the trial and error process through which life and consciousness progress. They, too, are part of our evolving community.

In the last several chapters I have been discussing our attitudes toward and relationships with people who seem other than ourselves. I have been contrasting mechanistic attitudes with more quantum ones. In the next chapter I want to extend this discussion to the nonhuman, to the world of nature and physical reality. I want to look again at our all-too-human notion that the nonhuman is the ultimate "other," a realm of being different from ourselves in both substance and principle, and at the social and ecological consequences of this notion.

# 10. The Self, the Vacuum, and the Citizen: Our New Pact with Nature

*The natural world is subject as well as object. The natural world is the maternal source of our being as earthlings and the life-giving nourishment of our physical, emotional, aesthetic, moral, and religious existence. The natural world is the larger sacred community to which we belong. To be alienated from this community is to become destitute in all that makes us human. To damage this community is to damage our own existence.*

Father Thomas Berry*

*There are no messiahs. . . . If we wait for a messiah, we wait; if we react to the challenges of present and future as responsible moral agents, as aspiring citizen pilgrims, then we act. When we act, a cumulative process unfolds, leaders emerge, and new horizons of realistic aspiration present themselves.*

Richard Falk†

The two quotes with which I have begun speak to two problems that increasingly dominate our attention and concern. There is, on the one hand, the growing ecological crisis that threatens the very survival of the planet and hence the lives of all who dwell here. Our lakes and rivers are polluted, often

*Dream of the Earth.*
†*Explorations at the Edge of Time.*

beyond their ability to sustain life. The air we breathe is foul and sometimes poisonous, our trees are dying from the effects of acid rain, the ozone layer's protective shield is rent, food additives and fertilizers poison the food chain, whole species are suffering annihilation, and so on in this sad and all too familiar litany of impending ecodisaster.

At the same time, we are nearly paralyzed by an increasing sense of personal impotence in the face of this and a whole host of other problems on an apparently global scale that nonetheless impinge on, sometimes even threaten, our daily lives and livelihood. What can any one of us as individuals do in the face of the world's ecological crisis, global economic turbulence, widespread poverty and inner-city decay, rising crime figures, Third World hunger, the spread of AIDS, the growing irrelevance of existing political structures, or the spiritual aridity of modern culture?

This sense of our common, personal impotence was put to me by my American editor as a challenge that she wondered whether this book could meet. "Your greatest problem," she wrote when the book was still in the planning stage, "will be to make any book on society seem relevant to your readers. It is possible to see what an individual can do to become a 'quantum *self*,' but what can any of us do about the enormous array of social problems that stand between us and a 'quantum *society*'?"[1]

I think the two problems, the problem posed by our global ecological crisis and the problem of personal impotence in the face of wider social issues, are really two aspects of the same, deeper problem. Both, I believe, issue from an underlying spiritual crisis that now grips almost the whole of Western culture.

We face a spiritual crisis not just because we have a less easy faith in the old gods and the old, established religions but because, along with this, we have lost our sense of what it is to be human. We have lost our sense of where the deep roots of our humanity lie, and hence we have lost touch with the source of our own efficacy as personal and moral agents. We no longer know what gives us the *authority* to act. We don't know from whence comes either mandate or capacity to act. At the same time, we are profoundly out of touch with the whole natural side of human being. We have little "lived" sense of nature and the physical world or of our place within it. For the first time in history, more than half of the world's human beings live in cities. We don't "naturally" see our humanity and our social concerns within a wider context that includes nature. We don't see that "[our] social institutions are damaged and threatened by many of the same forces that threaten the environment."[2]

I think that we can recapture both—both the source and meaning of our humanity and our authority to act—and some more positive relationship with the natural world. But to do so we need the transforming vision to see that the two are inextricably linked. We need the vision to see that the deepest roots of our humanity are to be found *within* the wider natural world. We need to recapture the natural within ourselves, and to see that *within such natural rootedness lies our empowerment to act.* Given where we are starting from, this is a "tall order."

The whole history of mainstream Western culture these past two thousand years has been a history of distancing human nature and the efficacy of human action from any association with matter and the natural world. Both main pillars of premodern culture, Platonic philosophy and the Christian Church, rest themselves on a radical distinction between mind, soul, or reason on the one side and body, nature, or experience on the other. Both stress that the "best," the "highest," or the "truest" of our human qualities are those that rise above or stand apart from nature and all her processes, particularly the natural processes of the human body. Nature is the Other against whom or in spite of whom we define our humanity.

For both the Platonist and the Christian, the source of human efficacy, too, lies outside the natural domain. The Platonist believes our authority to act issues from recalling our innate knowledge of the eternal Forms, the Christian that it issues from our immortal soul and its grounding in the (disembodied) Divine Being. Both believe this efficacy is diminished by any contact with the body or its physical world. "So long as we keep to the body and our soul is contaminated with this imperfection," warned Socrates, "we are lost to our pursuit of truth."[3] "Oh wretched man that I am," echoed St. Paul, "who shall deliver me from the body of this death? . . . So then with the mind I myself serve the law of God; but with the flesh the law of sin.''[4]

The Christian attitude toward nature itself ranged from seeing it as simply pagan to judging it actually vile or corrupt. The Protestant theologian John Calvin argued that the Fall of Man meant the Fall of Nature as well. Adam's sin, he claimed, had ''perverted the whole order of nature in heaven and on earth.''[5] A century later the great English poet John Milton described nature as filled with ''naked shame . . . sinful blame . . . Confounded, that her Maker's eyes/Should look so near upon her foul deformities.''[6] Whether pagan or vile, Genesis reminds us that nature is ours for the taking. ''Be fruitful and multiply, and replenish the earth, and subdue it: and have dominion over the fish of the sea, and over the fowl of the air, and over every living thing that moveth upon the earth.''[7] More than two millennia later Calvin repeated, ''The end for which all things were created was that none of the conveniences and necessities of life might be wanting to men.''[8]

Some of our contemporary sense of the human and of our relation to nature rests on fragments and vestiges of Christian teaching. We still tend to think of ourselves as the top of Creation's tree and of nature as being there to meet our needs. But for a great many people, the whole Christian worldview was undermined by the scientific and philosophical revolution of the seventeenth century. Descartes' famous doubt and his project to destroy the foundations of all his previous beliefs[9] ushered in the new Age of Reason.

For the new rational humanists, human beings were still to be ''the centre of creation and the measure of all things,''[10] but our exalted position followed not so much from possession of an immortal soul as from the capacity to use our minds clearly. It is through Reason that we know the difference between truth and falsehood, between good and evil, and through which we derive the will and the power to act. In Reason lies both the source of our humanity and its efficacy. But this modern sense of Reason was very narrow.

Where earlier thinkers had equated reasoning with the power to answer questions like ''What is a good life?'' or ''What is valuable?,'' the sense of Reason promoted by Descartes and the other seventeenth-century rationalists was more bound up with logic and mathematical truth. "Considering that among all those who have previously sought truth in the sciences," Descartes wrote, "mathematicians alone have been able to find some demonstrations, some certain and evident reasons. I had no doubt that I should begin where they did."[11] Earlier, an idea was judged rational by whether it made sense within a whole wider framework of ideas (within a worldview). To escape this "inner logic" of his previous beliefs, Descartes decided that all problems should be broken down into as many (isolated) parts as possible. He would believe only what he could observe to be true, or that which could be deduced clearly from first principles, like a geometric theorem from its axioms.

This whole reductionist and atomistic method got taken up by Newton and the other great mechanists to become the basis of our rational, scientific culture.* They applied it to the study of nature. Nature as the Other became nature as Object, nature as that which can (which must) be considered apart from the human, that which can be broken down, reduced, and measured. The old Platonic and Christian dualism between the soul and experience or between the mind and the body became the new scientific dualism between the (rational) Observer and the (nonsentient, or nonconscious) Observed. Rationalism and the scientific method became the new paradigm for all thinking, for all assessments of what counts as true or what counts as valuable. Thus we arrive in the eighteenth century at the philosopher David Hume's famous "Fork":

> If we take in our hand any volume—of divinity or school metaphysics, for instance—let us ask, *Does it contain any abstract reasoning concerning quantity or number?* No. *Does it contain any experimental reasoning concerning matter of fact and existence?* No. Commit it then to the flames, for it can contain nothing but sophistry and illusion.[12]

Hume's strictly limited criterion for what counts as "true" or "worthwhile" inspired the whole positivist streak in Western

---

*Descartes was not the only author of the new scientific method. Significant contributions were made by Francis Bacon and Galileo, among others.

thought and philosophy. But it leaves out a wide range of things and pursuits that we value. It leaves out all of religion, aesthetics, and morals. It even leaves out much of the reasoning that we do,[13] and certainly most of common sense. Our usual, everyday thought processes are intuitive and holistic, not scientific and piecemeal. Even our simple perceptions—something like recognizing a face or a style of writing—cannot be done rationally. Serial computers are notoriously bad at pattern recognition.

I think it symbolic of the whole modern age, of our narrow sense of Reason and the equally narrow sense of self that accompanies it, that Descartes had his first great vision while crouched deep inside an old Bavarian coal "stove." He was a young soldier stationed at Ulm in Germany and on a cold November day he sought refuge in the tiny brick room dominated by its iron coal stove at the center. After spending the whole day in deep and secluded reflection, he fell asleep and dreamed that he was destined to found a new unified science of nature based on mathematics. Where prophets of old had their visions on mountaintops or under open desert skies, the first prophet of modernity had his while sealed off from the elements and from all meaningful outside influence. The modernist self (and the modernist sense of the human) is similarly thrown back solely upon its own (introspective, mental) resources.

Reason and its scientific expression promised us a great deal. It promised to rid the human mind once and for all of prejudice and superstition. Through technology it promised to improve immensely the quality of life. And it was meant to make us free, to give us rational plans of action, a rational means to solving all our problems. It was meant to raise us above the animals, above the constraints of nature and the natural. And yet each of these promises has had its unforeseen dark side.

The promise to free us from superstition and prejudice has had the unwanted consequence of undermining much of what we have believed and valued about human being and the wider claims of spirituality. The human itself, like the natural world, has been reduced to the determined, the predictable, and the "scientific." Reductive theories of evolution and artificial intelligence have robbed us of our "soul." As the Benedictine monk Dom Bede Griffiths has said,

> There has been a reversal of human values, a spiritual breakdown, which has brought into play forces beyond the material and the human. The present crisis has been prepared by the whole system of science and philosophy, affecting religion and leading to atheism. This has released forces beyond the material and the human. As St. Paul says, 'we are not contending with flesh and blood but against the principalities, against the powers, against the world rulers of this present darkness.'[14]

Many of the admittedly simpleminded or preposterous claims of the old religion were shown to be "unreasonable," but Reason's narrow interpretation of what counts as true meant that nothing has taken their place. Prejudice, rather than being abated, has if anything increased through Reason's claim to singular truth. The utter rejection of any kind of spiritual truth by scientists like Richard Dawkins is sheer and unreasoned prejudice justified in the name of Reason. The moment that I believe my way is the only way I become unable to see others' ways as having any value. They are at best "wrong," and possibly "revisionist" or "treasonous." Dedicated disciples of Freud's "scientific psychology" (including Freud himself ) are often noted for their refusal to recognize the validity or the therapeutic worth of any other approaches to the self. The "reasoned" foundations of Marx's scientific approach to history led to the bloody repression of both political dissidence and religious expression. Mao's Cultural Revolution alone cost the lives of at least ten million.

Technology has indeed done much to improve the quality of life. Public health and sanitation improvements alone have doubled life expectancy among those whom they benefit. But technology has also given us the rape of the environment. It has given us pollution, nuclear weapons, and an uncontrolled birthrate. Its reduction of time spent working has exposed our impoverished sense of what it is meaningful to do with our leisure.

Reason as applied through technology and rational organization theory has become "instrumental reason," the reason that sees reason as an *instrument for solving problems,* the reason of "how" rather than of "why" or "what." Instrumental reason has indeed made us more free in some senses. It has freed us of many of nature's constraints. It has indeed liberated much of our working time, but it has also given us Weber's "iron cage" of bureaucracy, the numbing, value-neutral, person-discounting system of organizing human beings like the replaceable parts in a machine. Instrumental reason and the bureaucratic mind have taught us to treat nature and human beings alike as objects for use and manipulation. In the hands of an authoritarian state they have the power to enslave and degrade. At their most extreme, they have been employed to discover the most efficient means for annihilating people.

Hitler's Nazi hierarchy presented the Final Solution as a problem of logic and administration. The Reich wanted to rid itself of certain enemies. What would be the most rational and most efficient means for doing this? The gas chambers themselves were viewed not as instruments of death so much as the best technological solution to this problem. The reasoning process used to solve the problem took no account that the "obstructions" to be removed were human beings. They might just as well have been termites, or autumn leaves. That value-neutral, problem-solving mentality which underpinned it is what some see as the true horror of the Holocaust. It is somehow far more chilling and nightmarish than even the most virulent hatred ever could be. As sociologist Zygmunt Bauman has observed,

> It was the spirit of instrumental rationality, and its modern bureaucratic form of institutionalization, which had made Holocaust-style solutions not only possible, but eminently 'reasonable'. . . .[15]

Because of Reason's dark side, because of the intolerance, the failed promises of infinite progress, the failed promise of a rational solution to all our problems, because of the pollution, the bomb, the Holocaust, and a whole host of other "sour fruits" of Reason, we have lost faith in Reason. Beginning with Nietzsche's claim that Reason is just another myth, we have seen the gradual unfolding of the whole deconstructive postmodern assault on truth, the claim that all truth is relative, just a limited "perspective" or "inter-pretation." Deconstructive postmodern philosophers and sociol-ogists have directed their attack particularly against the claims of instrumental reason and scientific or technological "truth." The perspective of science, they say, has no privileged claim to truth. Science, too, is just another "narrative," another "fiction."[16] Rea-son itself has no value for its own sake. Many deconstructive post-modern philosophers have argued that it is just another instrument of power.

At the popular level, the deconstructive postmodern rebellion against Reason has led to a whole array of irrational practices and beliefs—the claims of some alternative medicines, crystal therapy and healing, reawakened interest in astrology, shamans, witch doc-tors, "nature religions," alleged sightings of the Virgin, and many other superstitions associated with the New Age or the new fun-damentalism. There may be some value in these various practices, for those who turn to them, but adopted as blind dogma they are just a further rebellion against reason. Sometimes their irrationality seems to mean more to their followers than their content. In so many ways reason has lost its (fragile) hold on the postmodern mind. At the same time we have lost faith in ourselves, in the fundamental basis of our humanity and our efficacy.

Since Nietzsche's assertion of an "end to Man" we have seen the slow but inexorable undoing of the whole "metaphysic" or worldview of Enlightenment humanism. Descartes' "thinking self" (*cogito ergo sum*), the self that has access through "clear and distinct ideas" to all fundamental truth, has evolved to become the empty "nothingness" of the existentialists or, more recently, the "decentered self" described by French psychoanalyst Jacques Lacan. This postmodern (deconstructionist), posthumanist self is grounded in nothing larger than its own conventions, associations, and interpretations. It is, in Sartre's words, "condemned to be free"—free from all "nature," all meaning, all commitment, and from any means to effective action. From some, this evokes a cry of pain. "In our current situation," asks American ecofeminist Charlene Spretnak, "where is reason? Where is coherence? Where is truth, beauty, a love of life? Where is even the most basic impulse for self-preservation? What have we become?"[17]

What we are left with as a result of our lost faith in Reason and in the human self grounded in Reason is "that there is no longer an authentic metaphysical, humanistic, or technological reality in which we can believe."[18] In the more personal words of a recent Iris Murdoch character, "We could lose our ordinary sense of the world as ultimate, our self-being, our responsible consciousness."[19] We could easily lose faith in our own future. This is the spiritual crisis of which I spoke a few pages back.

Yet, ironically, this spiritual crisis may have arisen just at the moment in the history of our planet when it is most needed. Western dualism, the whole history of the Western separation between human beings and the natural world, the Western preference for grounding human nature and human efficacy outside nature, even *against* nature, has brought us to our present moment of ecological crisis. So, too, has this crisis been exacerbated by the allied Western notion that human beings are "chosen," that we are creatures apart (ontologically different), that we are the center of things, the purpose for which all other things and creatures were made.

*I want to argue that we could not respond positively and creatively to the now desperate state of our environment from within the old Western worldview, be that Christian, humanist, or mechanist.*[20]

Neither, however, can we respond to our ecological crisis from a position of crippling despair (no worldview) nor from some romanticized vision of the past (a nostalgic worldview). Despair is the last word of the (deconstructive) postmodernists. For them there can be *no* human efficacy and thus we are powerless to act. Nostalgia is often the way of the Green Movement. For Greens, human efficacy very often lies in a return to the traditions of native peoples or of small rural communities. Action means abandonment—of modern technology, of large-scale industry, of big-city life, of notions of progress.[21]

*I want to outline a new way forward. I want to describe the foundations of a quantum worldview. This is a worldview that sees human beings as an integral part of nature's wider processes, a worldview that grounds both the possibility and the necessity for human action in an unfolding of those natural processes.*

## The Quantum Worldview

The quantum worldview is, in its very essence, an ecological worldview. But it rests on an ecological vision that can *use* rather than *abandon* science and technology, a vision that exalts and expands the foundations of human reason rather than undermining or limiting them. And it is a vision that takes us forward from where we are now, using the best of all we have achieved while relinquishing the attitudes that have made those achievements destructive.

The basis for the quantum worldview is twofold. At one level, there is simply the uncanny analogy between the structure and processes underlying quantum reality and the structure and processes associated with conscious mind. This analogy alone makes the insights of quantum physics a powerful model for our broader thinking about the nature of self and society. Quantum physics is, after all, a discovery or a language used by the human mind to describe physical reality. There is no good reason why that same language can't be used to describe human reality if the potential for doing so is there, and I believe it is—both metaphorically and, as I will suggest again in a moment, in fact.

We have already discussed several of the relevant insights from quantum physics that would feature in such a model, whether it is metaphorical or actual—the possibility of a both/and logic, an appreciation of the role played by pluralism in quantum physical processes, the necessity to see the natural world in holistic terms. But, for purposes of articulating a more ecological worldview, perhaps the most important insight of quantum physics is the mutually creative relationship between the observer and the observed in the quantum domain.

The quantum observer does not stand outside his observations. He does not see nature as an object. Rather, he *participates* in nature's unfolding. The observer is *part* of what he observes. This has very important implications for a more ecologically sound relationship between human beings and nature, between the new science and nature. It has the potential to give us new insight into how we might use our technology for "greener" ends.

But there may be a second, more fundamental basis for the

quantum worldview, one that takes the relationship between the observer and the observed onto a more profound level. I have argued this throughout the book so far. This more fundamental basis lies in the link I have suggested between the *actual* physics of human consciousness, the potential structure of human society, and the fundamental physics that underlies all else that is in the universe. If we have good reason to believe that the nature of mind, the nature of society, and the nature of nature are all one and the same thing, that all are linked by a common physics, we have then a firm foundation for grounding both human nature and human efficacy in the (fundamentally) natural within ourselves. This is the deeper logic behind the juxtaposition of the words "quantum" and "society" in the title of this book. And it is the more concrete reason for arguing that the language of quantum physics can be used to enrich our description of human reality.

In Chapter 3 I outlined why there are good reasons to think that conscious mind may arise from the peculiarly quantum characteristics of a structure called a "Bose-Einstein condensate" emergent from the brain's neural processes. This argument currently has the status of "grounded speculation." There are experimental indications that it is true, and its associated predictions are compatible with the latest discoveries of neurobiology. If it is true, it is important to understanding the whole physical history of consciousness and the place of the human mind within the wider scheme of things. If it remains just an analogy, we are still left with our powerful quantum model.

Bose-Einstein condensates are so called because they are made of *bosons*.* Bosons are one of only two *basic* sorts of "particles" that make up the whole universe. The other sort are called *fermions*. Fermions are particles that make up things. These include electrons, protons, and neutrons, the basic constituents of the atom. All the matter in the universe is made of fermions—my hand, the keyboard on which I am typing this script, trees, rivers, the earth itself, etc.

Bosons, on the other hand, are "particles of relationship." All

---

*Both bosons and the Bose-Einstein condensate take their name from the Indian physicist Satyendranath Bose, who first defined their characteristics.

the fundamental forces that bind the universe and its constituents together—the electromagnetic, the gravitational, the strong and weak nuclear—are made of bosons. Bosons include things like photons, gluons, and *gravitons,* if they exist (Table 10.1). Consciousness, if it is an emergent boson phenomenon, might be another such binding "force," the means by which the disparate data generated by experience are bound together in the unity of our conscious field or the related unities of our conceptual schema and "worldviews."

Fermions are essentially "antisocial" particles. They slightly repel one another and so always stay at a distance. This is why matter is solid, why it doesn't simply collapse (into all that space within the atom). Bosons, on the other hand, are essentially "social." They like clustering together. Their wave fronts can overlap and they get inside each others' boundaries.* This is why forces (boson fields) pull things together. And it is why Bose-Einstein condensates are the most unified structures in nature. The bosons of which *they* are made overlap their boundaries completely. They literally share an identity, become as just one aggregate particle. This is why laser beams are so coherent, and why a Bose-Einstein condensate could be the physical basis for the unity of consciousness.

It is through this tendency of bosons to cluster together, beginning wherever two bosons meet and leading at its most extreme to the total unity of a Bose-Einstein condensate, that we might trace the physical history of consciousness, and hence of the human, back to its origins in fundamental physical reality. There is an experiment that graphically illustrates these first origins. It is known to physicists as "the photon-bunching effect."

In this experiment a beam of photons ("particles" of light, which are bosons) of constant intensity is fired from a source toward a detecting screen. On grounds of common sense, we would expect photons to hit the screen one by one at regular intervals. But this is not what happens. Because bosons "can't resist" each other and thus have a tendency to cluster, the photons

---

*Fermions *can* get inside each others' boundaries somewhat, but never to the extent that they coalesce.

## TABLE 10.1 BOSONS AS BINDING "PARTICLES OF RELATIONSHIP" IN NATURE

| Force | What It Does | Boson |
|-------|-------------|-------|
| electromagnetic (the force relevant to everyday life as we perceive it) | binds electrons to atoms; responsible for some chemical bonds; present in all living tissue. Makes consciousness possible? | photons; virtual photons |
| weak nuclear | binds the nucleus | neutral z, ± w |
| strong nuclear | binds quarks together to make particles (3 quarks make a proton or a neutron) | gluons |
| gravitation | binds masses to each other (so holds the universe together) | gravitons? |

(Table taken from *The Quantum Self*)

## FIG. 10.1 "PHOTON BUNCHING": THE PHOTONS (PARTICLES OF LIGHT) BUNCH TOGETHER IN CLUSTERS RATHER THAN ARRIVING AT THE SCREEN ONE BY ONE AS THEY LEFT THEIR SOURCE.

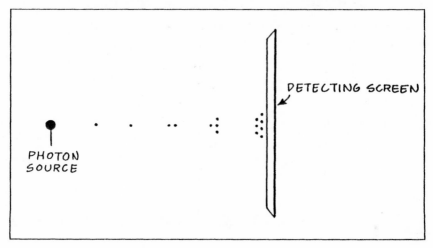

group themselves into "sticky" little clusters. They arrive at the screen at odd intervals, whole bunches at a time[22] (Fig. 10.1).

This photon bunching is the first small step, I would argue, in the gradual evolution of consciousness. The photons are not conscious, but their tendency to bunch is the origin of those much more complex structures (Bose-Einstein condensates) that might support consciousness in biological tissue. It is also the first stage in the evolving history of complex organization that later shows itself in the emergent complexity of quantum self-organizing systems, from simple ones like whirlpools in superfluids to more highly ordered ones like those found in living things. The coherent quantum structures found in things like yeast cells and bacteria as a result of Fröhlich's work* are such quantum (Prigogine-style) self-organizing systems. So, too, I have argued, is the physical basis of consciousness.

There is a whole new "metaphysic" of the human in this history of boson evolution. If the same tendency for two bosons to bunch together at the most basic level of early physical processes can be traced in unbroken sequence to the principles underlying the physical basis for conscious mind, we have traced the origins of the human back to primordial physical reality. We can see from this the possibility that the human mind, the self, carries within itself the whole history of the physical universe, just as the many layers of the evolving brain carry within themselves the whole history of biological life on this planet.[23] We also see in the world-view to which this would lead a radically different interpretation of the relation between mind and body and between human beings and (at least) all other living things.

If the same quantum structure (Bose-Einstein condensation) is responsible both for biological coherence in bodily tissue and for the coherence of consciousness arising from processes in brain tissue, there is no basis for any radical distinction between mind and body. Both are stuff of the same substance, though each, because of different biological structure, has a distinct (though integrated) function.

The same is true of the traditional Western distinction between

*See discussion in Chapter 3 of this book, or in *The Quantum Self*, Chapters 6 and 13.

human beings and other members of the animal kingdom. Though there is ample reason to expect a difference in the *kind* (structure and contents) and *degree* (complexity) of consciousness found in human beings and the "lower" animals, there is no reason whatever to suppose that a creature possessed of both a Bose-Einstein condensate and even the most simple neurological system would not possess *some* consciousness, however rudimentary. To a lesser, but still important extent, we share some common substance with *all* other living things, plant as well as animal, in that Bose-Einstein condensation has been found even in primitive yeast cells.*

There is no basis, then, in the quantum worldview for any ontological distinction between the human and the natural. This is a radical shift away from the whole earlier Western worldview. Quantum nature cannot be seen as the ultimate Other. It cannot be seen as Other in any meaningful sense. Indeed, the "quantum self" only makes historical sense, only fully comes into its own, within the context of the natural.

There is, in fact, a still deeper level to all this, a level that has to do not just with the history and nature of the human self but also with human efficacy, with the quantum source of our empowerment to act as personal and moral agents. This deeper level springs from the nature and function of what physicists call the "quantum vacuum." It springs especially from the physical relationship of individual existing selves to this vacuum, to the role played by individuals in expressing the vacuum. The discovery and the exact mathematical description of the quantum vacuum has revolutionized scientists' understanding of ultimate physical reality. I will argue throughout the remainder of this book that a knowledge of the vacuum, and of our relationship to it, is essential for a similar revolution in our understanding of human reality.

---

*This does not imply, however, that plants possess sentience, or "feelings" (emotions). Neurological structure of the kind found only in animals is probably a necessary if not a sufficient condition for that. Sentience probably requires input (perception), output (action or response), memory, and a capacity for processing data. See I. N. Marshall, 1989, p. 81.

## *The Vacuum as the "All"*

The vacuum is called a "vacuum" because it cannot be perceived or measured by us directly. When we *try* to do so, we are confronted by a "void," a background without features that therefore *seems* empty. Trying to look directly into the vacuum is a bit like trying to see the universe in the black background that surrounds the stars in a night sky. All we can see is blackness.

We can see particles and we can see waves, but as we already know, neither of these is primary or permanent. Quantum reality consists of an inaccessible wave/particle dualism, and the waves and particles themselves can always be transmuted one into the other. At high energies, one particle can transmute into other particles. At the level of perceived existence, everything has a kind of impermanence.

To make sense of this cosmic dance of temporary realities,* physicists needed to understand what lay beneath it. If particles and waves are only manifestations, what are they manifestations *of*? To answer this question a whole new branch of physics called quantum field theory was born. According to this new science, everything that exists, all the waves and particles that we can see and measure, literally, as in the Greek, *ex*-ist or "stand out from" an underlying sea of potential that physicists have named the vacuum. Waves and particles "stand out from" or "wave on" the underlying vacuum, just as waves undulate on the sea.[24]

Physicists' first motivation to look for something like the vacuum arose in response to Einstein's relativity theory. Einstein proved that the once famous "ether" did not exist. The universe is not filled with a jellylike substance. But in that case, since light can be a wave, what is it a wave *on* or *in*? The later discoveries of quantum particle physics raised the same kind of question. Since particles can appear and disappear at random, what do they emerge *from* and where do they go *to*?

Experiments in particle physics showed that existing particles,

---

*Particles and waves that are manifest, though impermanent, are not "illusions" as in Hindu philosophy. In quantum physics, the emphasis is on a temporary and shifting reality rather than on unreality.

as well as coming and going from "nowhere," are also slightly moved or deflected from their predicted paths as though there were something acting on them. The greatly expanded mathematical framework of the new quantum field theory predicted such effects and attributed their presence to an all-pervasive, underlying field of potential—the vacuum. The vacuum, though unseen and not directly measurable, exerts a subtle push on surface existence, like water pushing against things immersed in it.* It is as though all surface existing things are in constant interaction with a tenuous background of evanescent reality. This background reality, the vacuum, now replaces the material, jellylike ether. The universe is not "filled" with the vacuum. Rather, the universe is "written on" the vacuum, or emerges out of the vacuum.

We can see then that the quantum vacuum is very badly named, because it is not a vacuum in the usual sense. It is not empty. On the contrary, the vacuum spoken of by quantum physicists, like the Buddhist concept of *Sunyata,* or the Void, to which it is so similar, is replete with all potentiality. The Buddhists say of the Void that it is not Something and yet it is not Nothing. "To call it being is wrong, because only concrete things exist. To call it nonbeing is equally wrong. It is best to avoid all description. . . . It is the basis of all . . . the absolute, the truth that cannot be preached in words."[25] In the language of physics, the vacuum is the "ground state" (the "minimum energy" state) of *all* that is in the universe. "The vacuum, empty space, actually consists of particles and antiparticles being spontaneously created and annihilated. All the quanta that we have discovered or ever will discover are being created and destroyed in the Armageddon that is the vacuum."[26]

Metaphorically, as I have suggested, we can think of the vacuum as a vast sea and of all else that is—the stars, the earth, the trees, ourselves, and the particles of which we are made—as waves on that sea. Physicists call these "waves"—us, and all else that is—"excitations" or "fluctuations" of the vacuum. *We* are excited states of the vacuum.

---

*Experimentally, this is called the "Casimir Effect."

Over long distances the vacuum appears placid and smooth— like the ocean which appears quite smooth when we fly above it in a jet plane. But at the surface of the ocean, close to it in a small boat, the sea can be high and fluctuating with great waves. Similarly, the vacuum fluctuates with the creation and destruction of quanta if we look closely at it. . . . "The vacuum is all of physics."[27]

In more religious language, the vacuum is *the* All of everything.

The vacuum is also, as it happens, a Bose-Einstein condensate. This was first suggested to physicists by the uncanny similarity between the equations that described Bose-Einstein condensates and the equations that describe the interaction of particles with the vacuum. It became more clear still when the Nobel Prize—winning discovery of *w* and *z* particles was formulated. These particles were found to unify the electromagnetic and weak nuclear forces because they come from the same source as photons and are made heavy by interacting with some underlying field (the Higgs Field). This field is a major constituent of the vacuum and is itself a Bose-Einstein condensate. As one quantum physicist describes it, "We can imagine that the 'vacuum' in which we live is analogous to a 'weak superconductor.' "[28] (Superconductors are a common example of Bose-Einstein condensation.) It is a perfectly coherent boson field phenomenon.

Thus the vacuum has the same physical structure as human consciousness if it, too, emerges from a Bose-Einstein condensate. And just as excitations of the vacuum give rise to existence (the information content of the universe), so excitations of a laser beam (holograms) give rise to the information "written" on the beam and excitations (neural, sensory) of the brain's Bose-Einstein condensate give rise to thought, to the content of mind, the mind's information. There is, then, quite possibly a common physics linking human being and human thought to the ground state of "everything that ever existed or can exist"[29] in the universe. This is a very exciting idea, filled with wider implications.

It follows, for instance, that there are probably two different but equally fundamental ways in which each of us can be related to the vacuum, or to the ground of all Being. On the one hand, we are certainly related to it because we are *part* of it. Each one of us as an individual *is* an individual excitation of the vacuum, an individual being on the sea of Being. This relationship does not depend upon any theory of consciousness put forward by me—it is a straightforward conclusion of orthodox physics. It is, if you like, proven, or written into the equations of physics. But second, each of us may also be related to the vacuum because we *resemble* it. This would be the case if my theory of consciousness is correct, if human consciousness does have the structure of a Bose-Einstein condensate.

## Human Beings as the Active Agents of Creation

There is in both these ways that we can be related to the vacuum a whole new sense of finding human beings at the center of things. We are not *the* center, as in the old Western worldview, but we are *at* the center, ontologically part and parcel of everything around us. "Where man is in the world, of the world, in matter, of matter, he is not a stranger, but a friend, a member of the family, and an equal. . . ."[30] This is a vision more common to some of the great wisdom traditions of native peoples or to the ancient Greeks, but here it is derived from the latest insights of science.

It is very difficult not to see this link between the physics of ourselves and the physics of the vacuum in spiritual or mythic terms. If we were looking for "God" within physics, the vacuum would be the most appropriate place to look. As the underlying ground state of all that is, the vacuum has all the characteristics of the immanent God, or the Godhead, spoken of by mystics, the God within, the God who creates and discovers Himself through the unfolding existence of His creation.

In this new "myth" of the vacuum, all things that are, are expressions of the immanent God's being. All are precious and awe-inspiring, all "filled with a spark of the divine." This is similar to the view put forward by pantheists for millennia. We human beings, too, like other existing things, are expressions of this divine Being or Godhead. We, our conscious minds, our thoughts, our relationships, our daily lives, our daily activities and decisions, the intentions that we harbor and the reasons that we give for those intentions are all like "thoughts in the mind of God." We are among the instruments of His own unfolding potential.

But as creatures whose consciousness may *resemble* the physics of the vacuum, there is also a very real sense in which we are "created in God's image." This is a condition we would share with all other living things (all others in possession of a Bose-Einstein condensate), but we may after all bear some special responsibility in the order of creation. Because we may be possessed of *the most complex version of this physics in the universe,*\* we may, in a strictly physical sense, actually be at the vanguard of evolution. This physical, yet specifically human complexity takes the form of knowing and a capacity for self-reflection. As Richard Falk expresses it,

> The human species has a special coevolutionary capacity and responsibility. Unlike other species we are aware of our roles in the world and bear the burdens of awareness. . . . As humans, we can respond to the pain of the world by devoting our energies to various kinds of restorative action, building institutional forms and popular support for . . . a dramatic reorientation of behavior.[31]

This awareness and the responsibility it confers gives a special weight to our human agency, a special meaning to all our actions, and a real, physical importance to the constructs of human reason. This is the true, the awesome, significance of "being created in God's image." As in O'Shaughnessy's famous "Ode,"

> We are the music makers,
> We are the dreamers of dreams . . .

---

\*This refers to the discussion in Chapter 9 of human thought as a complex pattern of "phase differences" or "phase difference eigenstates." Because of our sophisticated neural structure, our Bose-Einstein condensate is likely to have the *most* complex pattern of "excitations" possible—of any creatures or structures that we know.

Yet we are the movers and shakers
Of the world forever it seems.[32]

I have expressed the connection between the physics of mind and the physics of the vacuum in these "spiritual" or poetic terms because in this form they have an uncanny, and possibly a nourishing, resonance with older, more familiar spiritual or wisdom traditions. Indeed, there is a similarly uncanny link between many older, mystical visions and recent scientific insight—the Creation story of Genesis and Big Bang theory, the mystical sense that "all is One" and quantum holism, or the notion of *Sunyata* (the Void) and the structure of the vacuum, for instance. This may even be explicable given that conscious mind seems to carry within itself the whole physical history of the cosmos. We may be destined by the physical origins of mind to retell the same cosmic story whatever the form or the historical period of our myths and legends. Each myth and legend, each great spiritual vision, may be a partial (a time-bound and culturally influenced) expression of one ultimate, underlying reality and our relation to it. Seeing the possible link between physics, myth, and vision in this way may be meaningful to some people. It is to me.

Whether we see the vacuum and our relation to it in such mythic terms, or whether we limit ourselves to discussing these things in more purely "scientific" language, the implications for the source of human efficacy and for questions of human moral responsibility remain deep and powerful. In whatever terms we put it, as existing things we are expressions of the vacuum's potential. Whatever we think or feel or do is an expression of that potential, has the effect of exciting the vacuum or, in other words, of influencing the ground state of existence itself. And there are no other such complex expressions of this potential beyond existing things like ourselves. There is, as the old saying goes, "nobody here but us chickens."

Thus we come full circle to the second of the thoughts with which I opened this chapter. "There are no messiahs." There is nobody here to act but us. There is nobody to act *for* us, nobody to "save" us. We, each one of us as individuals, are the active agents of creation. We are the active agents of a *literally physical* dialogue with the vacuum that affects both ourselves and every-thing else that is.* Perhaps the "Second Coming" for which so many people have waited is nothing else but this—the realization that *we* are that "coming." *We* are the messiah. The job of trans-forming ourselves and saving the world is down to us.

Perhaps we have now entered a more mature stage in the evolution of consciousness where our science now gives us the vision to see the responsibility that we shoulder. This is the so-bering implication of the new pact with nature we achieve through the quantum worldview. And it is a radically different implication from the old and arrogant "Hooray, we are at the top of creation's tree" mentality. In the quantum worldview we are fully *part* of creation and its processes, we derive our being and our efficacy from those processes, and we are truly effective in the long run only insofar as we remain in dialogue with them. We are not the end for which all else was created but the humble servants of a larger drama that needs us for its own unfolding. We are *called upon* to enter a more mature stage.

When we act as individuals, we do so through our daily lives and through our close or intimate relationships. But we have seen over and again throughout this book that we are not just individ-uals. Each of us is, at the heart of our being, embedded within—*defined by*—further relationships. Each of us is a citizen, a member of society. Each of us bears responsibility for our society, for the policies and practices and mores of our society. Each of us to a large extent creatively discovers him or herself through these social engagements.

This sense of social engagement gives an added and pressing dimension to any newfound sense we may acquire of our agent responsibilities. It gives further added dimension to any new sense of social or political *priorities* (like saving the environment, housing the homeless, founding meaningful communities, etc.) that might follow from things we have learned or can learn about a "quantum

---

*According to physicists, when a particle travels through the vacuum, it stirs up the emergence of other particles (other excitations) *which in turn affect the original particle's path.* Thus as we go about our daily lives, we initiate actions that have repercussions which in turn come back on us. This again is uncannily like the Eastern doctrine of *karma*.

society.'' To be good citizens, to become what Richard Falk calls ''citizen pilgrims,''[33] we must embed all these feelings and insights in *political* action. We must engage ourselves actively and directly in resistance to the social and political ills we see around us. We must, as Falk says, be willing to ''get our hands dirty.'' In the next two chapters, I will discuss how we might do so effectively, using the principles of quantum nature latent within us as our guide.

# 11. The Politics of Transformation I: The Loss of Meaning

*They tell us that it is heresy to suggest the superiority of some value, fantasy to believe in moral argument, slavery to submit to a judgment sounder than your own. The freedom of our day is a freedom to devote ourselves to any values we please, on the mere condition that we do not believe them to be true.*

Harvard Graduate Student*

In his famous study of our contemporary moral crisis, *After Virtue,* American philosopher Alasdair Macintyre opens with a disturbing science-fiction account of a society that has had all its scientists murdered and all its scientific knowledge lost. A group of would-be enlightened citizens comes along much later hoping to revive these lost treasures. They give the scattered and idle scientific instruments some new practical use, they employ the old scientific expressions in their everyday language, and they put together theories of how the world works from fragments of the old ideas. But none of this new "science" makes any sense. It bears no relation to the real thing that was destroyed. It is just gobbledygook. Yet no one realizes this—there is no way they could.

*Quoted in Bellah, *The Good Society,* p. 44.

Macintyre compares this science-fiction situation to the real-life state of morality today. "We possess indeed simulacra of morality," he says, "we continue to use many of the key expressions. But we have—very largely, if not entirely—lost our comprehension, both theoretical and practical, of morality."[1]

I believe this same disturbing scenario applies to our political life and to the meaning and higher moral sense that once underscored that life. We go through the motions of public life, we have endless policies for the governance of that life and spend countless sums on these policies, but in the end we are left with what a great many of us perceive as political gobbledygook. For many of us, the meaning has gone out of politics. The moral discourse that might once have inspired any one of us to public commitment has become the hollow sloganeering of politicians whose words and actions leave us untouched. Like the Harvard graduate student whose individual freedom has lost its meaning, we feel, as London's *Independent* newspaper recently expressed it, that our wider democratic freedoms suffer "a sense of spiritual vacuum or nullity."[2]

All over the Western democratic world there is a sense of disillusionment and boredom with much of the political process. The very low voter turnout figures in many countries' national and local elections testify to this. Even among those who do vote, there is very often the sense that one is voting for "the lesser of two evils," that there is no really meaningful, positive choice. Voters very often comment that the issues they deeply care about are not the issues being promoted by the professional politicians[3] or that, even if the politicians do make certain relevant promises, it is all "just words" or cynical maneuvering to win votes.

The whole language of politics has in many instances become debased. The right words are spoken, the ringing phrases that once summoned us to action are repeated, but no one responds. The words have lost their meaning, those who utter them their credibility. Most of us still hold on to some vague Western sense that we are free, but we no longer know what we are free *for* or how we can use that freedom in any effective way to better our own lives or the lives of others about whom we care. As Freudian sociologist Philip Rieff says, this leads to a sense of individual desperation. "There is no feeling more desperate than that of being free to choose, and yet without the specific compulsion of being chosen."[4] That is, without the specific "compulsion" of being asked meaningfully to serve or assist some cause larger than the self.

I believe strongly that we cannot separate these issues. We cannot separate the sense of personal meaning and the value of personal freedom from the wider sense of public meaning and our more all-embracing democratic freedoms. Nor, I believe, can we separate either of these from a deeper sense of spiritual meaning, from a sense of what we value, what we think is good, how we define the "good life" or a "good person," what we think our freedom is for, what we think *society* is for, wherein we think the deep roots and meaning of society lie. This more spiritual dimension to our shared public life, spiritual in a wholly nondeified, nonreligious sense, is what has been lost for some time in our Western political process. It is the dimension whose loss has slowly bled that process of its meaning, value, and efficacy, and that has exacerbated certain divisive trends in society that today have reached crisis proportion. Today's political crisis, I would argue then, is first and foremost a spiritual crisis.

Historically, the earliest roots of this spiritual-cum-political crisis can be traced back to the early Christian attitude toward the "state" (government of the day) and the necessary management of public life. When Jesus advised his followers to "render unto Caesar the things that are Caesar's and unto God the things that are God's," he introduced a split between spiritual life and political life that was foreign both to the Greeks and to the Jews.

The Greek notion of "virtue" had bound together a vision of "the good life" and "the good man" with the role of the citizen in the polis. The Hebrew concept of the Law embraced the notion of divine manifestation in daily life and in the ordering of worldly affairs. Christianity, on the other hand, drew a sharp distinction between this world and the next and between the spiritual qualities necessary to enter the next world and the worldly qualities needed to survive in this one. There have been periods in Christian history—the political dominance of the Church in the Middle Ages, the rise of Puritanism, today's "liberation theologies"—when

Christians have tried to wed the two, but on the whole Christian spirituality has remained otherworldly spirituality and has spurned the political life. Added to this today is the fact that, in most Western countries, organized Christianity is no longer a force that excites the public conscience or inspires the public imagination.

There is, of course, a very "postmodern" version of this Christian otherworldliness described by Richard Falk. This is "a cultural disposition evident in certain circles, especially prominent in California," to suppose that the individual is doing something worthwhile about the problems of society by meditating or adopting "the appropriate psychological stance" without ever engaging in real political action or becoming directly involved in real social problems. Several California groups or ideas that Falk mentions in this context are "the Aquarian conspiracy," "the Hunger Project," "World Beyond War," and recent publications of the "Ark Foundation."[5] In Britain, the group known as the "chattering classes" might be one such circle.

In the sixteenth century, secular forces perpetuated the spiritual-political split, but the emphasis was reversed. In his very influential guide for ruling politicians, *The Prince,* Machiavelli urged his sovereign to attend to the political while largely spurning the concerns of the spirit. Nothing good could come politically, Machiavelli argued, by a ruler's concerning himself too much with such things as generosity, compassion, or honoring his word. His prince, he advised, must learn how *not* to be virtuous, and to make use of this or not according to need. "The fact is that a man who wants to act virtuously in every way necessarily comes to grief among so many who are not virtuous."[6] The cynical manipulation of people and appearances, and a healthy resort to force, was far preferable to too much concern with being a good person.

All politicians who followed in Machiavelli's wake were not wholeheartedly Machiavellian in principle, but the notion that secular government should be a world apart from the power of priests or the wider concerns of the spiritual life stuck in most Western political models, indeed in most Western models for organization in general. Unfortunately, indeed disastrously as it has turned out, for Western political reality,* among these wider concerns that got excluded from the political sphere were the vast majority of the local and diverse meaning structures of the small groups and communities to which people had traditionally belonged.

As a young child in the early 1950s, I had some personal experience of these close-knit groups, though this is unusual for someone of my age. I was raised by my grandparents and thus, by fortunate accident, grew up in a social context belonging to a much earlier generation, one which reflected rural life in the early years of the twentieth century. We lived in a small country town, Neapolis, Ohio, in America's midwestern farm belt. The town's population was only two hundred adults, all of whom worked in the local pickle factory, in the few local shops, or in jobs servicing local farms. Everyone in the town knew everyone else, knew their children and kept an eye on them, knew their health problems, their marital difficulties, their small "scandals," their achievements, and their contributions to the community. Everyone knew who could be trusted, who pulled his or her weight, who needed support.

Both of my grandparents belonged to a number of small groups that met in the town. Both belonged to the "True Blue Society" of the local Methodist Church, the group for church members over fifty years old. My grandmother belonged to the local teachers' association and to the ladies' Grange Society, a group of town and farm wives who met to plan local farm-centered events and to raise money for local charity. My grandfather belonged to the Elks, and to the local branch of the Ohio Justices Association, a group formed by justices of the peace from various nearby towns. Nearby my great-grandfather was the ward chairman of the local Democratic party, and a small group of party members met regularly in his sitting room. We all belonged, of course, to our own large, extended family, to the circle of my grandparents' fourteen siblings, their children and grandchildren. All lived within a thirty-mile radius of our town and some subgroup of these met at least once every week to spend Sundays and important family occasions together.

---

*And for Western management theory.

My grandparents and I were seldom alone in our house except when we slept. There was always a flow of townspeople or relatives meeting in our sitting room, gathered at the large kitchen table, or grouped around the bed to which my grandfather was confined much of the time with heart trouble. Our house and our lives seemed always at the center or on the periphery of the extended lives of the various groups who passed through and stayed awhile. There was a constant, ongoing conversation (about local and national politics, about religion, about town events or town policies, about what the local grocery store should stock and how much it was charging) that sometimes reached high-pitched argument but was always animated. My memories and the meaning of my childhood are inseparable from memories of all these people, their activities and conversation.

The social life of Neapolis was typical of community life all over the world in preindustrial society. It is still typical of some, though increasingly rare, Western rural communities and of far more Third World communities. Archaeological evidence suggests that small groups organized within the larger structure of village-size communities have been an important unit of human social structure since the Stone Age. Most of those very early groups were family subunits of the community, organized around a communal home and a family "industry"—farming, crafts, smithing, etc. Earlier still in our history, we know that apes typically live in small social groups of fifty to one hundred members. Sociologists who find the small group the basic unit of human organization and basic vehicle for social change, and hence the basic unit for their own study, tell us that our small groups usually contain no more than twenty people. This was so of those I knew in childhood.

Historically, the small groups into which people have organized themselves have obvious features in common. Because they seldom have more than twenty members, all communication between members is direct and face-to-face. Relationships within the group are intimate rather than "legalistic." Members see their own identity, and often their own survival or at least their material well-being, bound up with that of the group and take personal pride in belonging. The groups consist of like-minded people with a common purpose who share certain values and have a shared sense of the group's reason for being. They are homogeneous. Most groups have a "spiritual" aspect, a collective meaning which is expressed through shared rituals or ritual objects and a shared narrative or mythology about past "heroes" (members of the group who had expressed the ideals of the group in some exemplary way). In my grandfather's Elks group these were men who had raised large sums of money for local charity through especially imaginative schemes.

In Neapolis, many of the town's small groups had a political dimension in the sense that their activities often spoke to or bore on the larger, organized life of the town. The meaning of my grandmother's Grange group was bound up with the meaning of life on the local farms and the practical necessities of farm education, distribution of farm goods, local ordinances relating to farm boundaries, farm vehicles and safety practices, regulations relating to grain storage, milk and egg production. My grandfather's local branch of the Ohio Justices Association concerned itself with codes of practice for local courtrooms, the ethics of local justices of the peace, the hours that constables would work, and the community role played by the justice and his premises.

All these small-scale political concerns (or at least senses of group purpose or group meaning) had the same dynamics as similar small, local groups throughout human history—sporting clubs, religious or cultural societies, professional associations, extended families, encounter groups, workingmen's associations, school clubs or classes, and so on. In premodern, or preindustrial society, politics and the community, politics and the local meanings and local values of local groups, were inseparable.

I want to argue that small groups and the role they can play in focusing identity, meaning, and local political activity are more than just a curious sociological or historical fact. I think their existence and character are a natural reflection of human nature, a necessary link between the structure of the human brain (and hence the nature of human consciousness) and the structure of human social and political organization.

I suggested very early on, when we were discussing our quantum model of consciousness, that the brain and consciousness are key to understanding the nature and potentials of society. Any style of thinking, perceiving, remembering, or behaving that is "natural" to the brain will often get expressed in an individual or a social lifestyle. But the brain has evolved and it has many layers of development. As new layers have formed through evolution, they have not replaced the older layers but rather have been added on top of them. Today, I believe that we can identify three distinctive organizational styles and their accompanying structures within the brain. The earliest is intuitive and associative and corresponds to the organizational style of traditional small groups. A second style is more rational and corresponds to larger-scale bureaucratic or scientific organizations. And the third, which may well form the bridge between the other two, is the quantum, which we will discuss in the next chapter.

Scientists distinguish two sorts of long-term learning and memory—habit memory ("implicit learning") and one-off memory ("explicit learning"). They differ in many ways.[7] Habit memory is based on repetition and practice, like a child learning to ride a bicycle or learning multiplication tables through constant repetition. It is a concrete, unreflective, associative knowing-how rather than a more logical knowing-that. It seems to be located all over the brain. It contrasts with our one-off memory of events or ideas.

One-off memory evolved more recently and involves a particular part of the brain (the hippocampus in the limbic system). It can deal with abstractions and representations—remembering a face we have seen only once or recalling the plotline of a novel we read once. One-off memory and the style of calculation that it makes possible works on principles far closer to those of our personal computers. It often deteriorates with age, whereas habit memory remains functional in the elderly. Older people have trouble remembering what was said to them yesterday, but they can still drive cars or ride bicycles if physically fit enough. Hence the expression "You can't teach an old dog new tricks."

I want to argue that the traditional small group organization in society reflects habit memory and the thinking patterns associated with it. In traditional small groups, customs are built up slowly through tradition (repetition) and are slow to change. Customs and traditions are remembered and embodied through habits—crafts, skills, rituals, songs, group gestures, etc. Small groups rely very heavily on these group habits for their continued existence and their sense of identity.[8]

Habit memory, the brain's model for traditional small groups, includes simple conditioned reflexes, but it can be much more subtle if large numbers of units (structures in the brain or individuals in society) are all capable of mutual readjustment. The recent and popular "parallel processing" model for thinking and learning mirrors these same dynamics.[9] In the brain's parallel processing, many groups of neurons* are interconnected; they exchange information and cooperate to act together as a unit. Learning is slow, via repetitive and gradual strengthening or weakening of the relevant interconnections. This conditioned-reflex-like neural process has been found in the single synapses (neural junctions) of simple organisms,[10] and is no doubt responsible for corresponding processes in human beings. The parallel system slowly acquires habits, discriminations, and skills by trial and error, but cannot reflect on what it is doing. It is widely believed that the parallel processing model fits simpler levels of human brain function but not the more complex and reflective ones.[11]

At the level of social organization, a simple example of parallel processing would be many people at a party sorting themselves out into groups of mutual interest during the evening, or people in a neighborhood each becoming regulars at one of the local pubs, bars, or social clubs. In each of these cases the individuals concerned would move around spontaneously until, by trial and error, they found a congenial group or atmosphere. They would not have made their choice by logical analysis, and might be hard put to express exactly why they had joined which particular group.

Small groups often carry out a great many of their decision-making processes by this kind of emergent, intuitive thinking rather than by logical, reflective consideration. Group members chat and discuss, sometimes argue; suggestions are made and a consensus emerges. This differs from a more bureaucratic "top-down" sort of organization, where people are assigned to particular groups for logical reasons, and, as we shall see, from a

---

*These groups are called "modules," and each module is thought to contain about 120,000 neurons.

"quantum" organization involving reflection as well as dialogue.

These first, early stages of thinking associated with habit memory are necessary to our whole conscious experience. They are the foundation on which it is built. But they can only take us so far. They learn slowly, through trial and error. They cannot deal with complex, rapidly changing, heterogeneous data—for instance the demands of life in a big city. They cannot construct rational plans for coping with unfamiliar situations, nor can they work with abstractions as is necessary when doing science or engineering.

In the same way, my simple life in a small, homogeneous group or a small village can only take me so far. Beyond that point it becomes constrictive, narrowing, claustrophobic. Just as simple perception belongs to the early development of the mind, life couched solely within the small group or community belongs to the "childhood" of our species. It belongs to a stage of life (actual childhood within a family group) or to a stage of social development (rural, premodern) that both the individual (as he or she becomes adult) or the culture (as it becomes complex) needs to grow beyond as the sole mode of expression. We value our life in small groups and the kind of tradition-bound, unreflective thinking that is associated with it, and these will always remain a valid and necessary aspect of our social experience. Politics at this early stage is a healthy grassroots politics, as in my grandparents' small town, but it cannot function effectively on its own as an adequate response to national or international political issues.

Similarly, in the study of organizations, theorists have noted that small groups function effectively in part because they develop a kind of "groupthink," a homogeneous response to people, situations, or decision making. But groupthink is also thought to be a problem. Some organization theorists "recognize that deviates, minority opinions, or unpopular views can serve a positive purpose in groups, since they cause the group to re-examine the problem and its alternatives."[12] In other words, diversity, and the conflicts that inevitably arise wherever there is diversity, is not easily handled by the small group. "Conflict can be a creative force if managed properly, but the groupthink phenomenon prevents this from happening, since conflict is never allowed to occur."[13]

In response to the complex demands of consciousness, indeed

perhaps even in response to the complex needs of more sophisticated human social life,[14] the brain developed further levels of
integration and more complex kinds of thinking. Socially, we are
struggling to do the same. The modern mind, beginning with Descartes and his doubt, wanted to break free of the limitations imposed by tradition-bound society and unreflective thought—from
groupthink. It wanted to break away from habit or revelation or
"hand-me-down" ideas.

The demands of life in big cities and heterogeneous cultures
require that we grow beyond our simple habits of thought and the
simple grassroots relationships and politics of our small, homogeneous communities. New political theories supposedly more
suited to a life lived reflectively through logic and reason, more
suited to life lived outside the constraints of local meaning, gripped
the modern imagination. In the West, the most potent of these
has been liberal individualism and the new theory of the "social
contract."

To the modern Western mind, meaning became something to
be pursued individually and in private, not through the political
process, the community, or in one's place of work.[15] This new
split between the public and the private mirrored Machiavelli's
split between the secular concerns of politics and the spiritual or
ethical concerns that princes must ignore. It also mirrored Jesus'
distinction between God and Caesar. In liberal individualism such
distinctions took the form of limiting the nature and scope of the
public sphere. It was embedded in the liberal idea of the "social
contract," in the whole wider notion that we contract away certain
specified powers to the political realm but we keep to ourselves
everything else. Social contract itself was inseparable from the
whole modern notion of the individual, of his supposed autonomy,
of what this new autonomous individual's identity consists of, and
of how we define the "self."

We saw in our earlier discussion, particularly the discussion
of liberalism and the liberal self in Chapter 5, that individualist
notions of freedom see the self as "prior to all its associations."
That is, the modern self is what sociologists would call a "disem-
bedded self." Our whole concept of identity is a *private* concept—
the self is rooted only in itself. I will believe to be true, as

Descartes put it, only what I can know clearly and distinctly with the power of my *own* mind. The natural extension of this is that I will hold to be precious, that I will value, only those things that have to do with *me*, with what *I* hold to be valuable.

Modern identity is not defined through relationship to any outside authority. Similarly, it is not defined through relationship to others in the community, to its culture, or to nature. It is not even defined in terms of its intimate relationships. The self is "prior" to all these. The liberal individual, in the words of philosopher Michael Sandel, is "a person wholly without character, without moral depth, for to have character is to know that I move in a history I neither summon nor command."[16] But this limited self, unlike the unreflective, tradition-bound self of the small group or local community, quite naturally finds nothing of itself in the political process, finds no motivation to be *concerned* with the political process, except insofar as that process can further the self's own selfish ends. Politics as it is now defined through contract theory no longer incorporates local or private meanings. These are *outside* the realm of the political. This new contractual politics is there for me to *use*, society is there to serve my purposes, to enable me to pursue my private dreams. My freedom is a *freedom from* all the constraints and coercions not specifically contracted away in my limited agreement with the state.

It is little wonder that a radical split between the public and the private and an emphasis on private meaning would lead to a loss of meaning at the political level. This is particularly true in liberal societies where stress is laid on the neutrality of the public sphere. Not only is it the case that my meanings are private, but also it is important that I keep them so, that I not "impose" them on others. Thus very little, if anything, that I hold personally to be valuable should impose itself on the public realm that I share with others. This principle has been upheld, for instance—with still considerable controversy—in the United States over the issue of prayer in public schools. My prayer might not be your prayer, so it is best we have *no* prayer.

The result, of course, is that one perennial cornerstone of the spiritual dimension is left out of the schools' daily life. A whole dimension of larger meaning is left out. This is despite evidence that more than half of all people privately reflect about such issues as the meaning of life, the meaning of God, or the meaning of suffering and follow this reflection with some attempt to live by their views.[17]

The ideal of liberal individualism and the social contract was that the rights and the autonomy of the individual should be safeguarded. Everything that was not explicitly contracted away should be left to the individual to dispose of as he or she wished. The philosophy of this was a reaction to the earlier tyranny and domination of kings and priests. The guiding purpose was to protect the individual from the state, to empower the individual and to leave him for the most part the source of his own authority, moral and spiritual. Liberal individualism "is impatient with the trappings of tradition, mystery, awe and intuition, and it summons all authority to the tribunal of individual reason to explain itself."[18] The best politics was meant to be a minimalist politics. But all this was to turn sour. The neutrality of the public sphere was to turn back on that very same individual with a vengeance never foreseen when the liberal ideal got turned on its head through the political and social side effects of modernity and the mechanistic worldview.

The rise of mechanism and the associated rise of industrial society had three almost immediate social consequences: bigness, rationalism, and bureaucracy. The new age of the machine meant that people rushed from their small villages and rural communities to congregate in big cities where the factories were located. The old community structures and local means for enacting the necessities of political life broke down. Either people ceased to organize themselves in small groups or these small groups themselves became marginalized. They ceased to be a vehicle through which people could discover their meanings and relate those meanings to the larger social process. New forms of organization were required that could deal with people in aggregate.

Bureaucracy was the form of organization devised to meet these needs. Bureaucracy was meant to implement Reason in its most efficient form. Its ethos was to eliminate injustices, favoritism, and corruption by treating everybody equally. But bureaucracy was the *raison d'être* of liberal individualism stood on its head.

In the first instance, bureaucracy and the rationalism that motivates it were natural extensions of the liberal individualist spirit. Where liberalism and the doctrine of social contract intended to free human beings from political and ecclesiastical tyranny, the rational spirit was intended to free us from more subtle tyrannies—the tyrannies of superstition, convention, imposed moralities and revealed truths, outgrown customs, etc. The rules of the new bureaucracies and scientific forms of management were meant to transcend such parochial concerns and the local differences that weigh people down. They were meant to bring a further freedom, but in fact they brought about a change of emphasis that had the opposite effect.

Where the liberal argued that all personal meanings *should* be private, as a positive way of protecting those meanings from outside interference, the rationalist spirit underpinning bureaucracy insisted that all personal meaning *had* to be private. The rules of bureaucracy are concerned only with a limited few things common to all, ways in which everyone *who is reasonable* might be expected to think, feel, or act in given circumstances. Anything that falls outside these rules and the contracts we sign agreeing to abide by them is outside the concern of the bureaucratic system. Small local groups with their private meanings might flourish, indeed liberal individualism encourages them to flourish, but they have no bearing on the system, no power as such. Their meanings are of no consequence to the general rules by which society is organized.

We can all think of many examples of this type of bureaucratic ordering in our lives and cultures. In Britain, for instance, the Road Safety Committee decided (for on the whole very good *reasons*) that all motorcycle riders should wear crash helmets. For most this caused few problems, but the neutral rule overlooked the custom among Sikh members of the Indian community to wear their long hair rolled up in turbans. This made wearing a helmet impossible. The wearing of turbans is a religious practice and of deep significance to Sikhs, yet this local meaning was not considered when the rules were made.

Again, in Western factories and large companies organized as bureaucracies, very little account is taken of employees' private needs, affiliations, or loyalties. The conditions of employment are laid down and everyone is expected to fulfill them. If Saturday overtime is required, so much the worse for the Orthodox Jew who regards Saturday as his Sabbath. If the company needs an employee to move to some distant city, so much the worse for all the social ties that bind him and his family to their home.

But this kind of rule-bound neutrality means that most of my *life* is outside the concern of the system—my hopes, my dreams, my fears, my passions, my idiosyncratic ways, the things that I value, the customs, traditions, and relationships that have meaning for me. In a rational bureaucratic system I can find none of these things reflected. The result is not, as intended, that I am free to pursue these personal meanings in private. Rather I am, in the whole public dimension of my life, robbed of these meanings.

Power and authority has slipped from my grasp to reside in the neutral bureaucratic authority that treats me as a replaceable part in the system. I am reduced, disempowered. I am a number or a type or a stage in the process of production. My meanings are deprived of their meaning. Nowhere can I see them reflected in or responded to by the system. When this system is my place of work, I feel alienated as a worker. When the system is the political system that runs my country, I feel alienated as a citizen. I cannot feel a personal relationship with nor personal involvement in a distant and bureaucratized political system that has nothing to do with the things that I care about. This is particularly true when the political system is embodied in professional politicians who are distant and who seem to have their own agenda.

The effect of this on the society of which I no longer feel myself to be a part was illustrated poignantly, and instructively, by a study done by a Danish sociologist working with a group of fishermen in a small Norwegian fishing village. In this case, the sociologist was called in to analyze why production was so low at a village factory manufacturing fish sticks. The main cause, he found, was that fish were never delivered on time according to any predictable schedule. The townspeople blamed the fishermen, whom they saw as a drunken and socially rootless lot, for the imminent loss of their only industry. The fishermen themselves were demoralized, took little pride in their role in the factory's production, and suffered an 800 percent annual rate of personnel replacement.

The sociologist found that the fishermen functioned well as

small groups on board their own ships. Nothing was wrong with the way they ran their ships or went about their fishing routines. But they had little sense of relationship to the town or to the factory for which they worked. This was because of a great change in their circumstances since the end of World War II.

Before the war, the fishermen had been settled and respected members of the community. They owned houses in the town and sent their children to school there. But the war had broken up the old social life of the town and left most of the old houses destroyed. A new population moved in, the fish stick factory was built, and the once free and independent fishermen were now seen as servants of the factory and its schedule. The new fishermen were unmarried and shunned by the social life of the town. Though they still functioned well as fishing units, their larger meaning as free members of the community had been taken away. The factory and the town were simply the neutral means through which they could earn a living.

In this case, there was a happy ending. The sociologist got the fishermen together with representatives of the factory, the mayor of the town, and representative citizens of the town in weekly meetings. Feelings, grievances, complaints, housing problems, banking needs, social and financial aspirations were discussed. Within two years, both the life of the town and the efficiency of the fishermen had been transformed. Factory production was markedly up, fishing crews now had only a 25 percent turnover, and most of the fishermen settled down in the town to married lives.[19]

The fishermen's tale ended happily because the meaning of their trade and the meaning of their small group coherence was successfully wedded into the larger meaning of the town of which they were now a part. They became part of the town's economic, social, and political life. But this is a rare outcome.

All over the Western world it is more the rule that individuals and small groups are not drawn into the larger social context. It is far more common that by a twisted but nonetheless inevitable path, the dream of the liberal Enlightenment has led in the West to the nightmare of bureaucracy and its loss of meaning. The liberal ideal of the unencumbered self, the privatized self with no roots in any tradition, community, or set of relationships, has led to the "organization man," the prefabricated, standardized, and rootless creature who finds himself nowhere in particular. It has led to "groups of fishermen who wander aimlessly through the town." This creature is without his traditional boundaries and has lost his "view of human life as ordered to a given end."[20]

The universal doctrine of the natural Rights of Man, rights that obtain regardless of culture or tradition or local sensibilities, has contributed to the evolution of bureaucracy's "iron cage," a hard and fast set of general rules that apply to all regardless of any particular personal characteristics or life circumstances. The liberal doctrine of the social contract, the doctrine that society should consist only of a set of minimal and value-neutral "functions of protecting all its citizens against violence, theft, and fraud and to the enforcement of contracts,"[21] has contributed to a society of minimal meaning and to the disenchantment of many individuals and small groups with that society. The bureaucratic view that relationships are simply legal contracts has invaded the more personal sphere. Concern for my family and neighbors or honesty in my business is reduced to a coldhearted morality of what is legally permissible. This was Jesus' charge against the Pharisees.

But all of this must produce a reaction. People cannot live for long without meaning, nor can they suffer forever a system that denies them identity. Both meaning and identity, as I have argued earlier, are built into the very physics of consciousness. They are built into the way the brain takes disparate bits of sensory data from the environment and binds them into a perceptual whole. At higher levels they are built into the way that the self-organizing system of consciousness weaves disparate bits of information into a worldview and into the way that essentially relational selves gather themselves into meaningful community.

Because it is a natural force, where this quest for meaning and identity is frustrated or denied in healthy ways, it may well be forced into pathological expression. We can see this at the personal, and even at the political, level in the rush into materialism, in the vain quest for meaning in the acquisition of "things" or wealth as empty symbols of self or identity or power. We can see it in the use of politics to further the interests of commercial or industrial cliques. ("What's good for business is good for America.") More potently still, we can see this pathology at work in the rise and predominance of the many factional groups and "isms" that blight today's political landscape.

One form the protest against meaninglessness in the public sphere may take is a flight into nostalgia. As philosopher Isaiah Berlin says, it may take the form "of a nostalgic longing for earlier times, when men were virtuous or happy or free, or dreams of a golden age in the future, or of a restoration of simplicity, natural humanity, the self-subsistent rural economy, in which man, no longer dependent on the whims of others, can recover moral (and physical) health."[22]

Such nostalgia is a dominant feature of Green politics. Many Greens hope to recover a lost realm of meaning through the wholesale dismemberment of modern society, ridding themselves not just of the bigness, the bureaucracy, and the pollution, both moral and physical, but also of much of the science that they believe brought it all about.

Nostalgia in itself is not necessarily destructive. It is antiprogressive and, I would argue, does nothing to further the progress of personal or social evolution. But simply looking back upon and hoping to recapture some imagined Arcadia brings little harm to anyone. It is when this nostalgia is linked to tribalism,* to the imagined glorious past (or the real or imagined repression) of some particular group—racial, ethnic, religious, class, sexual, or national—that it becomes a force that threatens to tear society apart. Such tribalism is the ultimate "ism" that represents the quest for meaning in its truly pathological form.

On the surface, tribalism answers to all our deeply felt needs for a shared, public meaning. As a member of the tribe I am not just an isolated individual. Nor do I feel impotent. I derive a shared history, a shared mythology, a shared glory (or suffering), a shared *existence* from my tribe. I can look at them and see myself reflected back and, indeed, enlarged. I get somewhat beyond myself, dis-

---

*"Tribalism" does not refer to the social groupings or customs of indigenous peoples. Its actual dictionary definition refers to a polarization between loyalty to one's own group and hostility toward, suspicion or dislike of other groups.

cover some further, tribal identity, that salves my sense of personal meaninglessness. Power and authority rest with my tribe and its chieftains, not with some distant and, to me, meaningless power structure elsewhere. And my tribe satisfies my need to go back, back to some simpler time when questions of identity were less complex, back to a time when I lay couched within the safe boundaries of my own group.

But tribalism is pernicious because it is a distortion. It tries to re-create the homogeneity and the meaning of the small group on a large scale. It fools itself into believing that the dynamics and attitudes of the small group are adequate as a response to larger and more complex society. Take, for example, the most virulent and most dangerous instance to occur on the world stage in the past century and a half, the form of tribalism that we call nationalism.

Nationalism, as Isaiah Berlin describes it, "is a pathological form of self-protective resistance . . . the inflamed desire of the insufficiently regarded to count for something among the cultures of the world."[23] It appears wherever people have felt themselves too cut off from their original meanings and small group identities, where their natural sense of group meaning and group identity has been threatened or shattered or made to feel of no consequence.

Nationalism caught the world unawares. None of the great historians or philosophers of the early nineteenth century saw it coming.[24] Before its rise, people were organized into states that included many nations, i.e., groups that shared a common language and set of customs and traditions. For instance the British state still comprises the four nations of Scotland, Wales, Northern Ireland, and England, and the former Soviet Union comprised Russia, the Baltic peoples, Georgia, and many others. These states fought over boundaries and sources of wealth, but they did not do so along "ethnic" lines. Wider patriotism expressed toward the conglomerate state was at least as important as loyalty to one's cultural in-group. But beginning with the demise of Napoleon's empire and continuing today after the collapse of the Soviet Union, Europe experienced a surge of narrower group loyalties to one's own "kind."

Nationalism has many roots or possible causes—each analyst

who writes about it today has his own favored theory. Isaiah Berlin attributes it to a reaction against liberal rationalism and scientific planners, both of which belittled and disregarded so many of the things that people in groups hold to be valuable—"the validity of the laws and customs and ancient ways of life—those impalpable and unanalyzable bonds that hold society together and alone preserve the moral health of states and individuals."[25] The scientific rationalist held much custom and tradition to be irrational vestiges of a primitive or repressive past. His new science (and its mechanistic thinking) would sweep away all these cobwebs, replace them with the neutral efficiency of bureaucratic organization.

Other writers attribute the nationalist surge to a reaction against the stresses of democracy in increasingly pluralist societies—an inability to sustain a sense of identity and meaning amidst a melee of many identities and many meanings.[26] Many attribute it to the humiliations or repression of colonialism, or to the new strain inflicted by globalization, both of which act to discount the local or traditional meanings of individual peoples.

Whatever the causes of nationalism, it is always a crisis of meaning and identity and always responds to secure these in an oversimplified and jingoist way that reduces all difference, all complexity to a ready formula that can be captured in the obvious rituals, slogans, or symbols of the small group. But what makes nationalism so politically dangerous and destructive, both internally and externally, is its use of the Other to define its own narrow boundaries.

Internally, the nationalist sees as Other all who do not conform to or agree with his own majority group. He "squeezes the nation into Nation . . .

> . . . by roaming hungrily through civil society and the state, harassing other particular language-games, viewing them as competitors and enemies to be banished or terrorized, injured or eaten alive, pretending all the while that [his] is the universal language-game whose validity is publicly unquestionable. . . ."[27]

He thinks he can force complex, heterogeneous society into his own deeply felt homogeneous model.

In Britain, a nationalist member of Mrs. Thatcher's government* invoked the famous "cricket test" for determining who was for Us and who was against Us. Indian, West-Indian, and Pakistani immigrants who cheered for their home countries' visiting cricket teams were against Us. They were Other, not "British." British "skin heads" and members of the neo-Nazi British nationalist party see these same people as fair game for demonstrations, beatings, and even murder. The same violence is used against immigrants and Jews in Germany, France, and Italy, all in the name of nationalism. In the former Yugoslavia, and in many territories of the now dismembered Soviet Union, "ethnic cleansing" has become the rule. If I am a nationalist, all my pains and woes, all my sense of feeling small or overlooked, is due to the machinations and scheming, the unwanted intrusion of the Other.

Hitler employed such logic in using the "Jewish conspiracy" as a means for uniting a divided and dispirited Germany. Neo-Nazi nationalists in today's Germany "liken Poles to hungry pigs and attribute shortages of bicycles to the Vietnamese and the lack of food to Jews."[28]

Externally, nationalism follows the same pattern, but now it is other nations or groups within nations who are the threatening Other that helps to define who we are—or rather what we are *not*. The United States for decades defined itself negatively as standing against Russia's "Evil Empire." British nationalism today is propped up by the alleged threat of European "Federalism." The Jews themselves have thrived as a people, in part, because their remarkable cohesiveness could draw strength from the constant, outside threat of anti-Semitism, the Other who is against them.[29]

The politics of nationalism, both internally and externally, is a politics of conflict and confrontation. I must keep the Other at bay, I must defend myself against him, defeat him. Therein lies my meaning, or at least its safeguard. It is very often more easy to blame my own shortcomings or lack of meaning on some scapegoat than to confront the sense of shame or guilt or helplessness that might accompany a more honest appraisal of my inner problems. The spiritual vacuity of the West only became clear when

*Norman Tebbit.

suddenly the obvious threat represented by the Soviet empire was removed. "It is as if the West needed an ideological opponent and a military threat to give meaning to its own institutions."[30] The realization that Judaism may have problems adjusting to the spiritual realities of modern life has only become clear now that less anti-Semitism has allowed a softening of old boundaries and a consequent flight into assimilation.[31]

I believe that nationalism is merely all our other contemporary "isms" writ large. Feminism, nativism, religious fundamentalism, the many ethnic-pride movements, lesbian and gay movements, all define their meanings narrowly in terms of the easily recognized, homogeneous tribe and negatively in terms of the Other who is different, who is wrong, who would oppress, ignore, threaten, or belittle them. Even larger, older, and more overtly political movements like trades unions and political parties have allowed themselves to become "isms." They define themselves in terms of narrow interest groups—sectional, professional or vocational, class, the single issue—and in opposition to Other interest groups. All fight their own corner against the Enemy Without and pounce readily upon the Enemy Within.

Thus blacks have their "Uncle Toms," left-wingers their "class enemies," fundamentalists their "heretics," feminists their "traitors," nativists their "assimilationists," and gays those who must be "outed." All this banding together of like-minded people is natural and, where people are oppressed, even healthy and necessary. But what makes this fragmentation into "tribes" or "isms" so destructive is the definition of one's own group *in opposition to the Other* and the casting of others in the role of *necessary* enemies.

The net result of all this modern crisis of meaning and the negative, tribalist quest for meaning that tries to fill the vacuum is that a significant percentage of the population are alienated from the political process altogether and the rest are caught up in the sterile and all too predictable politics of conflict and confrontation. My group, my interests, my meanings, my way are what must be fought for on the political stage. My advantage must be pursued, against the Other or despite the Other. As the Bellah team expressed it in *The Good Society,* claimant politics has come to overshadow civic politics.[32] That is, the politics of fighting my own corner has replaced a broader politics that seeks the common good of the whole.

All this is of course made stronger still by the liberal democratic tradition that politics shall be a pursuit of interests, that I best serve democracy by best serving my own interests, and that it is healthy for different interests to oppose each other (the politics of adversarial democracy). Faith in Adam Smith's "invisible hand" has not been justified by results.

When all citizens are either alienated or caught up in the narrow business of each fighting his or her own corner, no national or global consensus can arise. No true national or international interest can ever evolve. I think, particularly, of two instances of my-group-first nationalist political thinking that have marred international relations during the period I have been writing this book. In one, I remember Britain's prime minister flying off to a meeting of European ministers trying to negotiate further European integration with the words, "I'm going to negotiate the best deal for Britain." In the other, the French government was willing to undermine the interest of world trade and the global economy to fight for the narrow interests of French farmers in GATT trade negotiations. In neither case did the two countries in question think of any further community good or further community reality. Similar narrow corner-fighting by political parties typifies (and makes sterile) business in the British House of Commons every day.

When all meanings are private and/or exclusive, no larger meaning that is the shared meaning of all can ever emerge. In a contract society, meanings will always be private and in opposition. A contract society is an atomistic society. Where individuals band together in groups of parochial or restricted meaning it will always be a tribal society. The contract binds together the individuals or their tribes by only the most flimsy and most neutral of threads. The public sphere is seen as a mere facilitator of ends, a place where each will vie against the other to see that his ends are achieved. Where people seek these ends politically, the result must always be a sterile, ultimately a futile, politics of conflict and confrontation. This is a politics that fails to serve the purpose of politics.

Thus, at our present stage of social evolution we find ourselves in an impossible political bind. On the one hand the various forces of modernism have robbed public life of its meaning. On the other, in reaction, many people have turned to one or another form of tribalism in an attempt to recapture the kind of meaning found in older, more traditional societies and their shared public life.

Given only these two options, there is no positive way forward. We cannot live without meaning. We cannot forever tolerate the spiritual and political vacuity that follows in its absence. At the same time, our shared public life cannot long survive the factionalist pressures that would replace shared meaning. The many "isms" into which the American public consciousness has fragmented threaten all sense of social cohesion in that country and ultimately undermine the very meaning of being "an American." In other parts of the world, particularly in Europe and Africa, an ugly tide of nationalist conflicts leads only to ethnic war, displaced peoples, and death. Even the possibility of renewed nationalist conflict among the larger nations in Europe is no longer unthinkable.

To get beyond this bind and its destructive consequences, I believe that we need a whole new model for political organization. This new model, what I will argue I have good reason to call the model for a truly "quantum society," must do three important things. It must, in the first instance, incorporate the wisdom and truth, the value and the meaning of older, traditional social groupings, the small groups and their grassroots contribution to politics. It must also recognize that we no longer live in those traditional societies and understand what their limitations were, such that it was necessary to grow beyond them as our *only* form of sociopolitical organization. And finally, this new quantum model must suggest the natural roots of a whole further stage of yet unrealized social evolution, one which incorporates what is good of the past but within the context of a new, postmodern* political reality. It is to this further stage that I want to turn now.

---

*Here I use *postmodern* in its more all-inclusive sense, just to mean "that which comes after the modern."

# 12. The Politics of Transformation II: From Contract to Covenant

> Through most of empirically available human history, religion has played a vital role in providing the overarching canopy of symbols for the meaningful integration of society. The various meanings, values and beliefs operative in a society were ultimately "held together" in a comprehensive interpretation of reality that related human life to the cosmos as a whole.
>
> Peter Berger*

Politics is usually defined as the art or process of conciliation between different interests in society. For most of human history, a ground for such conciliation has been provided by a shared understanding of the common meaning and common "cosmic position" of the different groups or interests concerned. This common meaning was frequently, as sociologist Peter Berger says, associated with religion. The beliefs, symbols, and practices of a common religion offered a higher court of appeal in which conflicts could be resolved, a backdrop against which different parties could find some ultimate grounds for unity despite their many differences. Religion provided what Berger calls "the sacred canopy"[1]

*The Homeless Mind, p. 75.

under which politics could function successfully.

The union of politics and religion was at its strongest and most effective among those people who saw themselves bound together in society through the agency of a "covenant." Covenants are solemn and binding promises, usually with a sacred dimension, to fulfill certain mutual agreements. Because these agreements are sacred, they are grounded in a higher or further reality that transcends all local differences between the people concerned.

The most famous covenant in Western culture is that made between God and the Jews. In exchange for His own solemn promise to watch over and protect them as a people, the Jews in turn promised God to obey His Law and to perform certain daily rituals for the benefit of mankind. The meaning of Jewish existence and the day-to-day running of Jewish community politics are inseparable from this covenant. Muslims, too, have seen themselves as a covenant people, a community bound by certain divine duties and obligations and in return a community that shall receive certain divine rewards or punishments. Historically, Muslim communities were organized according to the prescriptions of the covenant, and disputes within the community were settled according to its possible interpretations. All Jews and all Muslims are brothers, stuff of the same substance, within their respective covenants. Christianity was supposed to be the "new covenant," but because of the split between spiritual and worldly concerns, this new covenant seldom took root as the binding force in the day-to-day political concerns of Christian communities. The Puritans were one obvious exception.

There have been other covenant peoples in history, and at the personal level many forms of private covenant that have the same meaning and function. Marriage used to be a covenant relationship (and still is for some) between a man and a woman who solemnly promised to abide by certain binding (and sacred) agreements. Politically, covenants were "the answer . . . to the age-old dilemma of civilization: how to maintain peace among a large and diverse population, perform the necessary social functions of cooperation and protection, and control individual attacks upon the security and property of others. . . ."[2] Privately, covenants are intended to provide a sacred, or at least a solemn, basis for sustaining a relationship through times of disagreement, conflict, or hardship.

Unlike mere contracts, which simply assign certain limited liabilities or responsibilities between agreeing parties, covenants usually call upon the whole wider meaning of the parties and their relationship, to each other and to some higher reality. A covenant can provide that "overarching canopy of symbols" that holds two people or a society together. Breaking a contract ends an agreement. Violating a covenant ends the relationship and the whole nexus of meaning that the covenant had defined.

Where the idea of a social contract binds those who live together in only the most limited way, leaving, as we have seen, most of their meanings outside the contract, the idea of society as a covenant is deeper, more all-embracing, more "spiritual." I believe that we need to recapture the spirit of a social covenant and to embed this spirit in our revitalized institutions. Were we to do so we might get beyond many of the problems posed by a loss of meaning in our limited contract societies, including our institutions' loss of public credibility. We might rediscover some common, even some sacred ground, for getting beyond the conflict and confrontation that has made our political life both sterile and dangerous. But even to suggest that some form of covenant might be desirable exposes at once the difficulty of finding grounds for one in today's postmodern political world.

## The Paradox of Pluralism

Postmodern reality is eclectic. It is plural. There are many truths, many values, many lifestyles, and many meanings that all vie for our attention and possibly for our loyalty. This eclecticism is the end result of the privatization of meaning, the fragmentation of shared, public meaning into what sociologists call separate "lifeworlds."

Religion itself has not escaped this privatization. It, too, has been marginalized, its once overarching values now a matter of private belief, even of private choice. The individual has now become what Peter Berger refers to as "conversion prone . . . Just as his identity is liable to fundamental transformations in the course of his career through society, so is his relation to the ultimate definitions of reality."[3] What Berger means is that, flooded with choice and aware that all these choices are available to us, we can move from one religious belief system to another, sometimes in an ephemeral or shallow way. Among the potpourri of religions now available to which any one of us might give our (limited) adherence, there is none that all of us can share as the underlying basis for our common social reality. This changes radically our attitude toward religion itself, but it also affects any possible relation between religion and politics. No organized religion as presently constituted can be the basis for a new social covenant. The eminent German theologian Hans Küng makes this same point in a recent book.[4]

Jung, and the post-Jungian psychoanalyst James Hillman,[5] tried to transform the privatization of religion into something positive by suggesting a religion or a mythology of the self, which all could share. This has given many individuals a deeper sense of meaning in their own lives, but it does little for the wider problem of providing society with a sacred canopy. "In this sense," as Philip Rieff noted, "Jung represents the uncertain and confused renewal of an effort towards personal knowledge that is also, at the same time, faith."[6]

In the neutralized public sphere of liberal contract societies, it was assumed that the state bureaucracy could step in to fill the vacuum left by the absence of shared meaning. The bureaucracy could arbitrate, like a neutral and impartial judge, between the conflicting rights and demands of the many private and sectional interests. One limited set of value-neutral rules for all would be enough to organize the separate life-worlds into a working whole. But bureaucracy is not enough. It is certainly no replacement for a covenant relationship between the parties whose interests it arbitrates. Bureaucracy, or instrumental reason, may be necessary in our large and complex society, but the bureaucracy itself must presuppose that there are underlying principles to what or why things are being organized as they are. These underlying principles are just what we lack.

These underlying principles are what I referred to earlier as the "Superego" dimension of society—our shared values, shared assumptions, shared practices, common purposes. All these are precisely what gets left out of individualist, or contract, accounts of society. Our rights are protected, we are free (autonomous), but we don't know what we are free for or how to relate to the freedom of others.

There is the further problem that bureaucracy's rules are imposed. They issue from a neutral rationality rather than from the local meanings and traditions of individual small groups. Thus they don't touch people where they are "at." When the bureaucracy arbitrates disagreements between the individuals or interests, the arbitration is accepted grudgingly. It may be deemed "fair," but the conflicting parties don't feel genuinely reconciled. An underlying tension may persist for years. This is true of most arbitrated settlements, compromises, or even votes—all are second-best solutions. They are "top-down" solutions to differences that could be resolved genuinely only from "bottom up," that is, from *within* (emergent from) the nexus of meaning around which people's lives are focused. The rules of bureaucracy are at best a crude tool for trying to define the formal "how" or "what" of a relationship between people. But in actual relationships, the *spirit* in which people do things is no less important than *what* they do.

To find real and lasting solutions to conflicts between different interest groups we need general principles that at once reflect the local meanings of small groups and at the same time provide some overarching meaning acceptable to all. Yet this overarching meaning is precisely what the privatization and pluralization of lifeworlds makes inaccessible. Thus we have what I would call the *paradox of pluralism*:

*Our need to accommodate pluralism has strengthened the privatization (and relativization) of meaning. Yet, it is only through discovering some collective, public meaning that we can learn to accommodate our pluralism.*

This paradox lies at the heart, indeed it *is* the heart, of our contemporary political dilemma. It exposes both the inadequacy and the sterility of modernity's brand of liberal individualism and the futility of all attempts to recover shared meaning through the denial of difference. The first can lead only to increased conflict and confrontation, the second to the repression of conflict through an imposed (and a false) homogeneity.

I think we can get beyond this paradox, but we can do so *only by discovering a renewed public meaning in the pluralism itself.* We must come to see that the full expression of difference is a part, a very important part, of the *raison d'être* of society itself, that commitment to the outright *celebration of pluralism* is the basis of the new social covenant that we seek.

We have seen that any covenant usually has its sacred dimension. We have also seen that we will never again find the common sacred dimension to today's complex society within the structure of any one of our existing organized religions. Each is committed to its own singular and private meaning. We do not now, nor, I believe will we ever again, live in a "Christian society," any more than Muslims can live in a "Muslim society" without ruthless repression and the denial of modernity, or Jews follow their ancient covenant in the same spirit while being members of wider society. But we can live in a sacred society. We can do so if only we can find our common roots in some meaningful dimension of reality that both undercuts and at the same time gives meaning to our differences. I believe we can find these roots in the quantum vision I have been outlining throughout this book.

### Our Quantum Covenant

We saw, particularly in Chapter 10 when discussing our relationship to nature, that both human nature and the human capacity to act as moral agents are grounded in our being expressions of underlying fundamental physical reality. We, like all existing things, are "excitations" of this underlying reality, which physicists call the quantum vacuum. We are "ripples" that stand out on the vacuum's sea of potentiality. We are a necessary part of (physical) nature's evolving reality. Beyond that, because the physical nature of our conscious minds may well resemble the actual physical structure of the vacuum (both are Bose-Einstein condensates), we are "created in God's image." We have a special reflective role to play in evolution's drama. We bear a special responsibility. The vacuum itself is, as I have said, the ultimate sea of potential. It is the ground state and the possibility of everything that is and can be. It may itself be a ground state of universal consciousness. If the vacuum is not itself conscious, it does at least have the potential to become conscious when its excitations take the right form, i.e., us and other conscious creatures. Someone raised to use the language of the Western Abrahamic traditions might ask: Is God conscious or is He only conscious through His creatures? Are we, in short, the thoughts in God's mind?

Because it contains *all* potentialities, the vacuum is also in its very essence plural. And it evolves through the expression and the meaning of those possibilities, through the creative dialogue between them. You, I, and everyone else are among these expressed possibilities. Our meanings, our cultures, our social groupings and political acts are among these possibilities. We are like the children of a common parent. A Christian would say "brothers and sisters in Christ" and a Jew "members of the People Israel," but the nature of the vacuum shows us that the Christ potential embraces us *all*, Christian and non-Christian alike, and that *everyone* belongs to the Community of Israel.

The dialogues among our various meanings, cultures, social groupings, etc., are the concrete steps in the unfolding process of the vacuum's evolution from potentiality to actuality. Thus there is a natural covenant between the vacuum and ourselves, between the source of all Being and ourselves. This is a covenant between the vacuum and our social and political reality. The nature of this covenant grounds all our meanings in a wider sea of shared potential meaning. It lays on us the responsibility to express as much of this potential—develop as much of its pluralism—as possible. This is, in fact, a more religious expression or outlet for the restless, modern urge for continual "progress" or "variety" that leaves us uncomfortable with traditional ways or the status quo but usually finds only distorted expression in continual technological or material progress. Materialism, at root, is really just a distorted quest for meaning.

This covenant between ourselves and the vacuum is *sacred* because it is about the ultimate meaning of our existence. If we call the vacuum by its other names, God or Being or *Sunyata* (the Void), this becomes more obvious. The vacuum is our common reality. It is the source of our being and of our being together. It is because the vacuum contains all potentiality and because the vacuum is within us that we carry the potentiality of the other within us. It is because of our inner dialogue with these potentialities that a dialogue between us as persons is then possible. Our covenant with the vacuum is also *political* precisely because it has pluralism at its heart, precisely because it calls upon us to conciliate between the different interests (different possibilities) in society and makes us aware that these all derive from a common and ultimate source.

Through this new "quantum" covenant all our private meanings acquire a new, shared public meaning. Each becomes a creative partner in the shared public enterprise of evolution. We, you and I, become partners in this enterprise—not *despite* our differences, but *because* of our differences. We need each other in order fully to become ourselves, to evolve that "dance" which is our common reality, and evolution needs us to be different in order fully to realize itself. Thus there is in this covenant both a horizontal and a vertical dimension. The horizontal dimension is the relation between us, between our individual characteristics and private meanings. The vertical dimension is our shared public relationship to the vacuum and to evolution, the relationship of our differences to their common source in the vacuum and the relationship of those unfolding differences to evolution's unfolding potential. Both are necessary dimensions of our political life (Fig. 12.1).

In the last chapter we saw how those brain processes associated with the basic perception of homogeneous data are similar to the processes by which homogeneous small groups or traditional communities structure their social life. The structure of the brain at that level seemed to act as a model, or as an "archetype," for the structuring and the functioning of the small groups. At this point I want to turn to more highly developed brain structure, to those brain processes that deal with the integration of complex, heterogeneous data.

I believe that an understanding of these higher brain processes will give us a natural bridge between the covenantal nature of postmodern society and the actual political structures and processes we might evolve to honor that covenant. I think these higher brain processes can serve as a concrete model for our transformed

**FIG. 12.1 THERE ARE BOTH A HORIZONTAL AND A VERTICAL DIMENSION TO OUR COVENANT WITH THE VACUUM, OR WITH THE UNDERLYING, FUNDAMENTAL, AND PLURALISTIC SOURCE OF THINGS.**

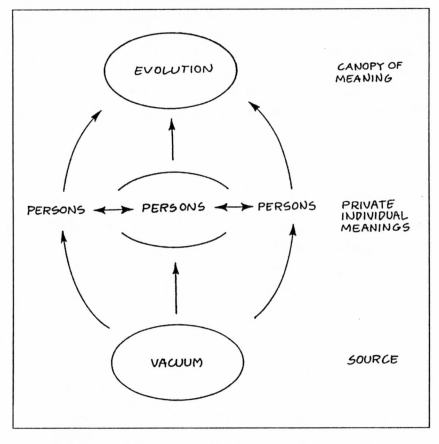

political processes. This is particularly true of those processes having to do· with getting beyond conflict and confrontation as the norm in our political life.

When the brain first receives data about a perceptual event, the data is transmitted along the various neural pathways to the different regions equipped to analyze particular elements of it. Those neural modules (groups of 120,000 neurons) that bear on the different features of one perceptual object then oscillate in unison to give a complete picture of the whole object. This is the simple level of perception of homogeneous data that underlies homogeneous small group organization. But this level of brain activity cannot accommodate difference. If there are any unexpected heterogeneous elements in the perceptual data, things that can't be handled through habit, conditioned reflex, or simple trial and error, the brain has to go into a "higher gear." It has to bring reflective consciousness to bear on the situation. Reflective consciousness is associated with the act of paying attention, consciously noting something and noting that I have to do something about it.

When something does call the brain to attention, for instance suddenly seeing a ball roll out from between parked cars while I am driving, a whole higher process of neural activity begins. As a first step, all the heterogeneous data bearing on the new situation are channeled to one region of the brain, the limbic system. This, despite being smaller and less developed than the more complex cerebral cortex, is the brain's center for integration. Indeed, the cerebral cortex is a more recently evolved outgrowth of the lower brain. It specializes in more detailed but less holistic analysis. The limbic system is like a forum or parliament or symposium where people gather to exchange their various points of view. All the diverse information from the perceptual field is gathered together and held together for three to six seconds (the "specious present").

A quantum theorist would argue that at this stage the data is looked at "holistically," held together in both space and time through quantum superpositions. During this time all the heterogeneous data is compared and contrasted and associations are seen. In my example, the association between rolling balls and pursuing children becomes clear. The data is also stored in the one-off memory system, the kind of memory that we saw is flexible and spontaneous as opposed to the slow and repetition-dependent memory of habit formation. One-off memory is often called "conscious memory," or "explicit memory."[7] It is the memory with which we recognize a new face or remember the day's events rather than the memory with which we slowly acquire a new skill like driving or roller skating. The two function from different parts of the brain.

If associations between the diverse bits of data being held together are easily seen, the brain has done its work. An almost instant synthesis weds the diversity into an emergent meaningful whole and the brain agrees on a plan of action. The social equivalent would be if all the people gathered in the forum or parliament, though arriving with different points of view, saw quickly that they had common cause, that their different points of view were really in harmony. They would quickly reach a consensus about group action. In both the brain and in society this quick resolution of difference (seeing the common thread emergent from the difference) is desirable, but it very frequently fails to happen.

When the brain "notices" that it cannot make immediate sense of all this diverse data, it begins a two-stage process of doing so through reflection. Socially, it would be as though the parties with diverse points of view agreed to some process of conciliation. Having looked at the material holistically and seen that no common thread emerges, the brain first "lets go" of the initial set of categories (the initial language) it used to tackle the data and begins to analyze the data anew by "deconstructing" it. It breaks the data down into simpler, more primary categories until it reaches a stage where some commonality underlies the differences. It looks at the data in a broader context by calling up further associations and memories. Then, as a second stage, the brain resynthesizes the data to build up a meaningful emergent picture.[8]

A concrete example of all this would be my noticing that a friend is behaving completely out of character. He or she is suddenly rude or does something dishonest. As a first reaction, I would put my previous impression of the friend on "hold." Then I would look for explanations of the uncharacteristic behavior— is he or she drunk, or upset? In the end a new picture would emerge. My friend is someone who sometimes can behave in ways different from normal or, more radically, my friend is not actually the sort of person that I had thought.

The very same process of deconstruction and resynthesis is similar to that described by some creative artists, who say that they have to "rid" themselves of old, habitual ways of seeing and learn to see the world in a new way. As Henri Matisse described the process, "Everything we see in ordinary life undergoes, more or less, the deformation brought about by acquired habits. . . . The effort necessary to get out of this rut requires a sort of courage; and that courage is indispensable to the artist who must see things as though he saw them for the first time; you have to see the whole of life as you did when a child. . . . ''[9]

In going through stages one and two, deconstruction and re-synthesis (which may take anything from a split second to weeks), the brain may have to learn a new language and invent new concepts. It finds the appropriate unifying concept (and accompanying plan of action) in a moment of "insight." This learning of new languages and new concepts and the moment of insight cannot be accounted for in mechanistic (AI) models of brain function. Serial computers can't work outside given languages (programs) and they cannot invent new concepts. Parallel computers don't use concepts at all, though they develop skills. But in a quantum model of consciousness, the formation of new concepts is the expected out-come of diverse quantum superpositions (diverse possible inter-pretations) "collapsing" toward a coherent theme in keeping with the process of evolution.*

The two stages by which the brain reflects on and digests het-erogeneous data to arrive at a coherent picture have both a hori-zontal and a vertical dimension. The horizontal dimension is the associations made between diverse pieces of the data as different neural modules oscillate in unison. That is, the various pieces of data relate to each other. The vertical dimension is the relation of the data across time. The one-off memory system allows the data to be compared to and integrated with past experience, while the process of collapse toward coherence (in the quantum model) links the data to the future through evolution. I described similar hor-izontal and vertical dimensions to our covenant with the vacuum— the horizontal being our relationship to each other, the vertical our relationship to evolution and to the vacuum.

This process by which the reflective brain deconstructs and then resynthesizes heterogeneous data to arrive at a meaningful

---

*This was the theory of collapse of the quantum wave function toward "possible states (eigen-states) of phase difference" that I discussed at length in Chapter 7. According to that theory, superpositions of diverse data, of diverse possibility, will tend to resolve themselves ("collapse") into states of greatest coherence, i.e. greatest meaning, because there is a universal principle of evolution operating at the level of fundamental physics.

whole reminds me very much of the "dialogue" process described by the late David Bohm. Being both a quantum physicist and at the same time passionately interested in the dynamics of society, Bohm was both sensitive to the role of pluralism in quantum reality and concerned with the problems and opportunities of pluralism in society at large. He was particularly concerned with how people of diverse opinion and diverse culture could get beyond conflict and confrontation to unleash the creative potential latent in their differences. "A free form of dialogue," he said,

> . . . may well be one of the most effective ways of investigating the crisis which faces society, and indeed the whole of human nature and consciousness today. Moreover, it may turn out that such a form of free exchange of ideas and information is of fundamental relevance for transforming culture and freeing it of destructive misinformation, so that creativity can be liberated.[10]

As I quote this I am reminded of my first date with my husband, and how a failure to engage in a free exchange of information and points of view got us both into an unwanted and temporarily disabling state of "destructive misinformation." We went to a Chinese restaurant and he asked me if I would like wine with my meal. I *assumed* that if he was asking, *he* must want wine, and I tried to go along with this assumed preference by agreeing. For his part, *he* assumed that as I had agreed I must prefer wine, and to be polite decided to drink it with me. We drank two bottles with our meal, both felt unpleasantly drunk, and only some months later were confident enough with each other to discover that each of us always prefers to drink Chinese tea with Oriental food. If we had exchanged that information in the first place, we would have seen that we agreed and would both have had a more pleasant evening. Even if we didn't agree, one or the other of us might have broadened our horizons and learned something.

Bohm's dialogue process itself calls upon its participants to go through the same two stages of deconstruction and resynthesis that the brain (the limbic system) uses to integrate its heterogeneous perceptual data. In dialogue, the deconstructive stage is described as a "letting go" or a "suspension" of one's own point of view as the only point of view. There must be a willingness to put one's own alongside others' points of view as one of many to be compared, contrasted, and considered. It is essential to the process, Bohm says, that one "hold many points of view in suspension" within one's own consciousness.[11] This, I believe, is like the brain holding many pieces of disparate information in superposition during the "specious present."

During this stage of the dialogue, one's own point of view becomes available for analysis along with those of others. The meanings and underlying assumptions of all the points of view can be analyzed, their cultural presuppositions and assumptions thus exposed, and their grip on consciousness loosened. "Such a thoroughgoing suspension of tacit individual and cultural infrastructures, in the context of full attention to their contents, frees the mind to move in quite new ways. . . . The rigid infrastructures begin to die away."[12]

Bohm believed that as people go through this process, much of the emotional charge surrounding a rigidly held point of view would be diffused, making it easier for participants genuinely to listen to one another. In the brain, it is a sense of frustration that is released—the mind stops trying to make sense of the data in the old terms (in the old language, with the original concepts). It frees itself to begin the process of reconstructing the data through the creation of new concepts and categories.

Once the participants in a dialogue have let go of clinging to their own points of view and the process of deconstruction is complete, the second stage begins the resynthesis. People discover they can listen to each other in a new way, that there is some common ground to be discovered. "When the rigid, tacit infrastructure is loosened, the mind begins to move in a new order."[13] This new order is a whole new, emergent level of consciousness in which the participants get beyond the fragmented state of individual consciousness to a shared pool of meaning and value, to a common purpose or understanding. They see that their original points of view in their original form clash, but if looked at in a new way "give rise to a unity in plurality."[14]

One obvious example familiar to most of us where such a

dialogue process has won through to at least a partial resynthesis is in the relationship between orthodox and alternative medicine. Once, the advocates of each could view the other only with the greatest suspicion and hostility. Each had its own tradition and set of assumptions. Neither listened to the other's point of view and they spent all their time vilifying each other. But as a result of public concern and interest, and the creation of many forums for discussion, slowly an exchange of ideas and a willingness to look at the actual evidence afresh has taken place. The result is a now somewhat more cooperative relationship where some forms of alternative treatment—osteopathy, acupuncture, meditation, and perhaps homeopathy—are recognized as valid and helpful. In combination with more orthodox treatments their inclusion in a patient's care now offers a broader basis for care than was previously available.

The truly revolutionary nature of dialogue can be appreciated by contrasting its processes of deconstruction and resynthesis with the confrontational processes in the more familiar, adversarial kind of debate or negotiation that presently underscores most of our democratic institutions. In adversarial debate, I do not "let go" of or suspend my own point of view. On the contrary, I sharpen it. I try to imagine the other person's point of view or set of arguments for the sole purpose of being better able to refute them. I do not place my point of view alongside the other's and regard them jointly as of equal weight. Rather, I fight for my point of view. I try to *persuade* the other, to win him over and convert him to my point of view.

In adversarial democracy, I judge my freedom and my efficacy by the degree of opportunity I am given to promote or defend my point of view. A good free-for-all of give-and-take debate is taken as a sign of a healthy democracy. This view argues that "justice is strife," to use the words of liberal philosopher Stuart Hampshire. "In the political and public discussion the adversary principle is a necessity, arising from the declared facts of conflict."[15] Similarly, some organization theorists argue that it is a "fundamental fact of human behaviour" that people in organizations will always pursue their own goals. Politics, in such organization models, is seen as the art of defending one's own corner—by fair means or foul.[16]

The liberal individualist sees politics as necessarily ridden with conflict, and thus as necessarily adversarial, because he has taken his model from the clashing billiard balls of mechanistic physics. The mechanistic organization theorist sees conflict as best resolved through the use of *power*.[17]

But just as in Newtonian physics the billiard balls (atoms) never change internally when they meet, so in such an adversarial political model, neither people nor their positions change very much through debate. People become entrenched rather than evolving. In the quantum model, by contrast, we have seen that two quantum systems merge when they meet. They get inside each others' boundaries and each acquires a new, further level of identity. In the brain, whole new neural pathways are laid down in response to experience. The brain is constantly changing internally, constantly evolving, as it responds to its environment and to its own responses to that environment. "The matter of the mind interacts with itself at all times."[18] In dialogue the participants' points of view evolve.

In adversarial debate, I employ deconstruction against the other's point of view, and he against mine. But neither of us deconstructs our own position. In dialogue, as in the brain, all points of view are mutually deconstructed. In adversarial debate I do not invent new concepts and new categories, jointly arrived at with the other, through which we find some common ground. Rather, I cling desperately to my concepts and hope that the other will come to think in my way. Compromise is always the best outcome that can be hoped for. Compromise, rather than consensus, is in fact the adversarial democratic ideal. According to Stuart Hampshire, "For the individual also, as for society, compromise . . . is both the normal and the most desirable condition of the soul for a creature whose desires and emotions are often ambivalent and always in conflict with each other."[19] But in compromise, rather than new ideas emerging and getting articulated, old ideas are slightly fudged to soften or disguise the points of conflict.

In adversarial debate I always see the other as *other*, as my opponent. I set out to defend myself against him or to best him. I "win" or "lose" a debate. In dialogue the other is my partner or fellow explorer in a mutually creative enterprise of discovery.

"In dialogue," as David Bohm says, "everybody wins."[20] To-gether we "win through" to a new, mutually welcomed position. In debate we are like boxers throwing our best punches and hoping the other hits the canvas. In dialogue we are like dancers exploring (and creatively discovering) our common score.

I have participated in such a dialogue process. I wrote about it briefly in Chapter 8 when I described the Athens symposium I attended where 150 people from different disciplines, different cultural backgrounds, and different religions met in an attempt to draft some common proposals for a new science of consciousness. Each of us, as I mentioned, came to the symposium with a "private language," a private set of assumptions, a certain amount of cul-tural conditioning, and a fairly fixed notion of what we each thought any new science of consciousness should entail. We found it difficult to understand each other and initially felt both frustra-tion and hostility when asked to listen to each others' points of view. But we did over the course of five days learn to listen.

I realize now that we learned to listen by going through pre-cisely the dialogue process described by Bohm, precisely the process used by the brain to integrate diverse perceptual data. We did not fall into a pattern of debate or sustained confrontation. We did not try to convince each other or win each other over. We listened, and through that listening a dynamic of its own de-veloped. The end result of that symposium was a group spirit and group coherence stronger than any I have ever known, yet it was a "dance with many dancers," a group of individuals who had found an emergent reality drawing our differences into a mean-ingful whole. The symposium's oral report took the form of a So-cratic dialogue that gave voice to all the main points of view—a "unity in plurality." The end result of "dialogue" in the brain is the unity of a perceptual field that contains many diverse ele-ments, not the suppression or manipulation of some data in favor of others as in the adversarial arena.

It was no accident that our symposium report was presented as a Socratic dialogue. The dialogue process described by Bohm goes back at least as far as Socrates. He used it to get young Athenians to break down their private assumptions and individual perspectives and to get them beyond disagreement to a common perception of truth. His skill at getting participants to realize that their own many different instances of something like "love" were each aspects of an underlying, more inclusive concept of love on which they could all agree was a classic use of the technique. (I think it was also classic "quantum reasoning.")

According to contemporary Greek philosopher Emilios Bouratinos, the whole of ancient Greek democracy before about 550 B.C. rested on a similar dialogue process. The citizens of Athens would gather in the *agora* (the market) and discuss whatever concerned them. They discussed problems with an open mind and their conversation continued until some resolution could be found. No votes were taken because the whole spirit of the meeting was to reach common agreement. "The practice of taking votes," says Bouratinos, "actually spelt the end of true democracy. It arose with the advent of the Rhetoriticians, who were advocates paid to argue positions and to put cases in the form of adversarial debate, as in contemporary courts of law." Bouratinos himself advocates cultivation of a "quantum attitude" to recapture democracy's lost inheritance.[21]

In our own times, the dialogue process has been used extensively by psychotherapists in this century* as an important tool for conflict resolution, though Bohm describes it with particular clarity. As I mentioned earlier, Gestalt therapists use it not just to resolve conflict between people but also to resolve inner conflict, getting individuals to engage in dialogue with themselves, between the different "voices" represented in their psyches. Bohm himself credits much of his own insight about dialogue to a London psychiatrist, Patrick de Mare.[22]

I think the uncanny resemblance between the dialogue process and the brain's own processes for integrating heterogeneous perceptual data is another example of nature providing us with a model and a basis for our own social behavior. The brain's higher processes of reflective consciousness give us something like a "dialogue archetype" or "complex group archetype" on which we can draw when trying to resolve our own differences and to make

---

*Particularly by Carl Rogers and the Gestalt therapists. See Carl Rogers, 1969, and Fritz Perls, 1969. Both give innumerable examples of dialogue.

use of our own diversities in a creative way. I also think that the dialogue process, and the kind of philosophy and commitment needed to make it work, could become the basis for a radically new political philosophy and accompanying political structures. This philosophy and commitment both have their roots in what I have described as our covenant with the vacuum.

In the brain, all the many pieces of disparate perceptual data are held together in the limbic system, the seat of reflective consciousness. If the brain has a quantum dimension as I have been suggesting, all this data takes the form of excitations (patterns) on the limbic system's coherent quantum field, on its "Bose-Einstein condensate." Each piece of data excites an oscillation of the underlying Bose-Einstein substrate. It is this feature, that all the data are excitations, or expressions, of a common substrate, that allows heterogeneous data to be held in superposition, or to be scanned holistically. The data are all "entangled" in the way that quantum excitations are. Their boundaries are overlapped and they share an identity. They are stuff of each others' substance.

Similarly, in society, each of us, as ex-isting things, is an excitation of the quantum vacuum (the ultimate Bose-Einstein condensate). Each of our coherent conscious patterns, our conceptual schema, our systems of value, our aspirations and fears, and our cultural attributes are excitations of the quantum vacuum.* As such we, too, are all entangled, stuff of each others' substance. This is the horizontal dimension I spoke of to our covenant with the vacuum—our relationship to each other. We are all aspects of the same larger whole, "faces" of the underlying "God" or, as Euripides expresses it at the end of the *Bacchae*, where "many are the shapes of things divine." This relationship to each other gives us one primary motivation, and our basic capacity, to engage in dialogue. It imbues our dialogue with a mutual respect and mutual cooperation, a philosophy of mutual belonging, that is not possible in adversarial politics.

---

*This is true whether or not consciousness itself is quantum mechanical in origin.

## Evolution's Dimension

But there is another dimension to our covenant that gives us further motivation to engage in dialogue. This is the vertical dimension, or our relationship to evolution. We have seen time and again throughout this book that evolution, at whatever level of reality it is taking place, feeds on variety. Evolution cannot happen without variation, without plurality; nor does evolution proceed without that plurality constantly combining to give new, emergent forms. Perceptual evolution requires that the various bits of heterogeneous data recombine (resynthesize) into new holistic patterns. Social evolution requires that different points of view, different ideas, different ways of life, and different traditions recombine into larger, more complex emergent wholes. Evolution imbues dialogue with a philosophy of mutual need and a commitment to the other's way or to the other's point of view as a necessary part of my own further way. This takes us beyond both adversarial politics and a politics of mere tolerance to a politics of partnership.

Politics as dialogue means politics as a perpetual ongoing conversation between as many different voices as possible. Politics as dialogue means more even than simply "the art of reconciling differences." It means politics as the art of creatively discovering new emergent realities *through the meeting of differences*. Among these emergent realities will be both political priorities and social and political structures.

Our old political priorities and existing social and political institutions are at present undergoing a process of deconstruction. This is one defining feature of postmodernity. As sociologist Philip Rieff says, "The systematic hunting down of all settled convictions represents the anti-cultural predicate upon which [postmodern] personality is being organized."[23]

Existing institutions and priorities are being scrutinized with a detached (or a bitter) eye and broken down into their constituent assumptions. These assumptions, or old bases for support, are increasingly found wanting. We are discovering that "the emperor has no clothes." Thus in Britain the central institutions of the judiciary, the monarchy, the Church, Parliament, the BBC, the schools, the main political parties, and many more have all come under some shadow of doubt or uncertainty. As one major newspaper put it, "The foundations wobble."[24] This same deconstructive process is happening in America and across Europe. The old verities no longer seem obviously true, the old institutions no longer so effective. As I suggested in Chapter 1, there is something deeply radical about our emerging social and political consciousness.

This deconstructive process exposes two things. First, it exposes the fact that the forces of postmodernity are such that we can no longer meet social and political necessity with the old categories, the old "language," the old institutions. This is like the brain realizing it cannot analyze the heterogeneous perceptual data in existing terms. But second, it shows us that new categories, a new language, and transformed institutions cannot be found through the agency of our old political philosophy or our existing political process.

The present situation calls upon us to follow the brain's model and resynthesize or reconstruct the data. But the political philosophy of social contract, the philosophy of liberal individualism, cannot do this. Individualism, with its private spheres of meaning, its adversarial politics of conflict and confrontation, and its neutralization (and despiritualization) of the public sphere can never give us the basis for dialogue. It can never give us the relation to each other or the sense of shared meaning necessary to conduct a dialogue in the right spirit. Still less, of course, can the fragmentation of the public sphere into conflicting factions give us this shared meaning or sense of common being. Beyond that, the bureaucracy that necessarily accompanies atomistic individualism as the sole binding force in society is too rigid to allow new forms, new categories, or new "languages" to emerge.

A more flexible, a more responsive, and a more dialogue-oriented political process would require new political structures that can themselves embody the dialogue, or quantum, ideal. Our existing political structures embody the adversarial, mechanistic ideal. In Britain, confrontation is built into the very design of Parliament as well as into its daily function. The government and opposition parties sit on opposite sides of the chamber, the opposition opposes *whatever* the government proposes, and the whip system whereby members of Parliament must vote according to party allegiance requires that most debate and most voting is conducted strictly along opposing party lines. It is very rare for something fresh or creative ever to happen. Members of Parliament come to the debate with entrenched positions and for the most part adhere to them without waver throughout. This is one reason British politics is so boring.

In every Western country party politics is built on the adversarial model. Politicians try to score points off one another, they try to best or outmaneuver one another, at very best they try to *persuade* one another.

But there is another factor that makes our existing political process mechanistic and unresponsive. That is the concentration of most power and decision making at the center, vested in increasingly few institutions and/or professional politicians who are distant from the daily lives of the people or small groups over whom they wield power and whom they are, comically, supposed to represent. In Britain, all power radiates outward from Westminster, in America from Washington. In Britain this process has gone so far that many commentators now claim Parliament is no longer representative or democratic. Washington is perceived to be peopled by distant (and often corrupt) bureaucrats. If we were to adopt dialogue as the central focus of our political process, the structures that embody it would have to be radically different, less centralized, less "professional," i.e., not entirely peopled by career politicians. Recently, as pointed out by David Osborne and Ted Gaebler, moves toward decentralization and dialogue have had many local successes.[25]

In the brain, the integration of heterogeneous data happens all over the limbic system, at different levels and in different centers. The hypothalamus deals with the data at the level of instinct, the hippocampus stores one-off memories, the amygdala deals with emotional response, and dozens of other centers deal with things like focusing attention or receiving input from all sensory areas. All these centers are in touch with other areas of the brain (Fig. 12.2). "This whole morphology," according to the description of neurobiologist Gerald Edelman, "interacts at many levels, from atoms up to muscles. The intricacy and numerosity of brain con-

FIG. 12.2 ALL THE BRAIN CENTERS ARE IN TOUCH WITH EACH
OTHER AND WITH THE CEREBRAL CORTEX. THE LIMBIC SYSTEM
CONSISTS OF CENTERS IN MIDDLE CIRCLE. (DIAGRAM IS VERY
SIMPLIFIED—THERE ARE *MANY* OTHER BRAIN CENTERS.)

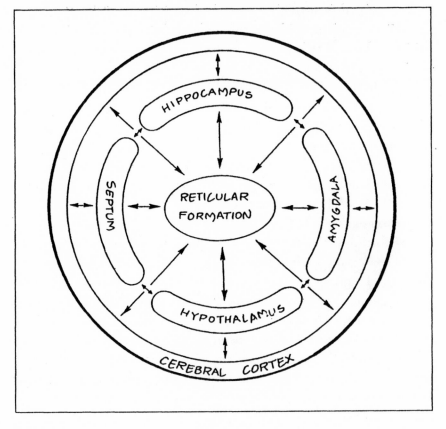

nections are extraordinary. The maps that 'speak' back and forth
have statistical as well as precise features. Furthermore, the matter
of the mind interacts with itself at all times.''[26] In the quantum
brain model all these connections are further enhanced by the
existence of nonlocal correlations across the different areas.

If we take the brain, especially the quantum brain, as our model
for how to realize our fullest social potential, a politics of dialogue
would stress the importance of many levels and many centers of
creative conversation. People would have to come together as
individuals, as small groups and societies and neighborhood associations, in think tanks, the universities, the trade unions, in local

and regional government, in parliaments and international organizations, and in a myriad other forums of the sort that presently exist or could be created. In our existing, highly centralized political system, most of the more local of these have been emasculated or marginalized.

The people who meet in these forums would have to do so as "independents," that is with an open mind, as ready participants in a creative dialogue rather than as advocates of some adversarial stance. This is not wholly incompatible with party politics. Political parties can be important foci of the more homogeneous, small-group-level meanings and values, and these must have a voice in the political process. But it would require a different attitude toward the primacy of "party loyalty." One's deepest political loyalty would be to open dialogue and to emergent, shared meaning, and to the underlying covenant of which that dialogue and meaning are expressions.

Historically, the first stages of many cultural creations—the early Greeks, the early Christian Church, the rise of modern science, the founding of America—were associated with many such open dialogues, before ossification set in. This "bottom up" approach has created new cultural, political, and spiritual realities. We have again today the same deep need to engage in this process.

The "independent" of whom I speak must have the characteristics of Richard Falk's "citizen pilgrim." The pilgrim is on a sacred mission in search of "a better country," a more "spiritual" one. He does not think existing realities are the only thing possible. The pilgrim "is not bound by any sense of duty to carry out the destructive [or the divisive] missions of a given territorial state [or a given political interest group] to which he or she owes temporary secular allegiance."[27] His true allegiance is to his own sense of higher responsibility.

All the many voices from all the many conversations would have eventually to feel themselves heard in the further conversations of representatives in more complex, more all-encompassing forums. The arts and the media should play the role of reflecting on and synthesizing these conversations. Parliament itself would have to reflect, indeed to amplify, not just the plurality of concern and opinion abroad in the land but also the emergent meaning resulting from the many-leveled and ongoing conversation. Parliament would have to become a forum like the limbic system, where old categories (old habits, conditioned ways of thinking) are deconstructed and new ones resynthesized. This could never happen within the present system of adversarial politics.

Political priorities, like the brain's plans of action, should also be emergent rather than imposed—emergent, flexible, and open to change in response to circumstances. Some very broad political priorities of any quantum society are implicit in our covenant with the vacuum. These include obvious things like reducing the causes of violence and any expressions of prejudice, avoiding domestic or international conflict, showing a regard for life, protecting the environment, increasing the availability and the quality of educational and training opportunities, and ensuring that no members of society live in abject poverty or in states of homelessness. But it is against the whole spirit of dialogue that we should write a new decalogue for our new society. It is far more powerful and far more representative of both our meanings and our potential that priorities emerge afresh from our ongoing conversation, and from the shared concerns and shared values that it represents. Keeping that conversation going, creating the multilayered and multicentered structures that make it possible, should perhaps be our highest political priority.

Of course, there are occasions when it is important to act in a more mechanized way, to impose priorities or plans of action. In times of crisis or emergency, or in situations where partners to a dialogue simply *can't* reach consensus, a government (or other authority) might have to impose actions or solutions. But any such imposition usually results in splits or stress that must be healed afterward by further dialogue. In Britain today, there is still endless public discussion about the wisdom of entering the Falklands War or the Gulf War, though both were imposed as emergency measures at the time. Such postmortems are one form that healing dialogue can take. In the same way, individuals often need to talk through stressful events or shocks for a long time afterward. This is an important element of trauma therapy.

It is very unlikely that the kind of transformed political process, and its accompanying transformed structures, that I have been suggesting will ever be initiated from the top. Those who hold the reins of power in the old system have little motivation to change things. Change must come from the grass roots, from those of us who see that the old system no longer meets our needs. The citizen pilgrim is ultimately the individual, me, and what I can do. But as Richard Falk warns, this does not imply a simplistic call for "power to the people"[28] or a rush to the barricades. It is a call to revolution, but it is a revolution that *must* begin with me. Unless the people, unless I, transform my personal attitudes, unless I attach my loyalty to the spirit of dialogue and to the "unity within plurality" that it embodies, nothing of significance will change. Many revolutions have led to "dark ages" or to successor regimes that were no better than the ones they overthrew. So the first step is the inner transformation of myself.

On a quantum view of the self, the self is deeply rooted within the fundamental processes of nature and evolution and deeply "entangled" (bound up in its being, its identity) with others. The self is the source of all transformation. All of us, as Jung said, "are not only the passive witnesses of our age and its sufferers, but also its makers. We make our own epoch."[29]

Each of us, as Jung realized and as a quantum view of consciousness would explain, may find in his or her own depths the collective yearnings, the collective fantasies, and the collective potential of the whole human race.* To find one's self at *this* level is to find all of humanity's potential for good and evil. To grow in one's self, at this level, is inseparable from helping others, from being more in tune with them, from helping them to grow. To transform one's self, at this level, is to transform others, to transform reality.

Psychotherapists find that they cannot easily understand or help patients whose conflicts closely resemble their own. Hence the requirement that therapists have had some form of therapy themselves. A therapist is best at helping with problems that he or she has learned to resolve—the capacity for resolution is then passed on to the patient like an act of "grace."

---

*Jung found a concrete example of this when large numbers of his patients began presenting dream images of a violent and barbaric nature a few years before the Nazi takeover of Germany.

In the same way, ordinary people usually tolerate least those faults in others that they have themselves. Even if this intolerance is not expressed, it is usually passed on in a felt way to the others. I must realize that my own upsets and conflicts are the psychological equivalent of an "infection" that I can pass on to those around me. Others do the same to me, but it evades my responsibility merely to ask that others change for my sake. I can do my best to deal with my own upsets, whatever their source. "The buck stops here." To realize all this is the basis for a renewed religious attitude.

But as I said in the last chapter, even such inner transformation as this is not enough. If I want to be a citizen pilgrim in the new quantum society, I must *do* something. I must become actively engaged in the political process. I must do this on two levels.

On the inner level, I must work on myself, educate myself, make myself aware of the issues and the forces at large in my society. I can do this through private reflection, through reading the papers, keeping up with events, attending the theater, involving my consciousness with the wider world around me, sorting out my values and priorities, realizing that social and political issues are issues for me.

On the outer level, I must become active in that wider world I share with others. I must make some interpersonal effort through conversation with family and friends, through "good works," through work, through membership in some group or association. I must get actively involved in the wider political process at some level that suits my abilities. If there is not a group that I can join that expresses my priorities, I might found one. But above all, in whatever I do, at whatever level I act, I must *listen,* I must engage in dialogue and do whatever I can to encourage dialogue between others.

*Every time I try to understand another person's point of view it is a small religious act. It is also a small political act.*

Similarly, government and public institutions have a spiritual, not just a political, responsibility to make room for dialogue, to encourage it, and to make clear that it is itself a spiritual process, the basis of our deeper, shared meaning (our covenant).

Chapter 13   Should be
required reading
OUT LOUD — stopping,
and going and discussing —
for every highschool
graduate!

---

Challenge:

make a list (alone) of
all the names of T.V.
shows that are about
FAMILY. Which ones are
traditional and which ones
are "quantum"?

    ... you can include
"Hazel," "Mr. Roberts,"
"Mr. Peepers" "Our
Miss Brooks," "Life with
Father," "One Man's Family"
and "Family Ties." But not
"The Honeymooners," no kids,
But you may not include
"Sex in the City," "South Park"
or "Dallas."

# 13. Reinventing the Family

*The very meaning of the family as we have known it has become
problematic. The poignant predicaments of the contemporary
family are in large part the most immediate result of the changes
and tensions in our larger society . . . of the fact that we no
longer understand the moral meanings of our institutions. . . .
We need not to "return to the traditional family" but to
understand what the family as a vital institution today would
really be like.*

The Good Society*

Last evening I had a small argument with my husband. We had both been working too hard to finish
this book, there had been tension all day, and finally
there was an exchange of insults and accusations. By
the end of supper, I had had enough. I stormed out
of the house and went to good friends for a calming
cup of coffee. I returned an hour later feeling much
better, almost ready to laugh about it all with my
husband, but when I went upstairs to kiss our children goodnight I found them both in a terrible state.
They were lying awake in their beds, convulsed in
tears and trembling.

"We thought you had left us," our son sobbed.

"We thought you were never coming back,"
echoed his sister.

---

*Robert N. Bellah et al., pp. 45 and 49.

I hugged them both and said they were silly. "Mummies never leave their children," I said.

"Oh yes they do!" was the joint and instant response. They then took turns telling me about all the children at their school who had either mothers or fathers leave home because of divorce. I knew the statistics. I read all the time in the papers about the divorce rate and the large numbers of children now living with a single parent, but I had never thought about it so personally nor realized what trauma this was causing to our own children. Since starting school, at least, our children have never simply taken it for granted that we will stay together as a family.

We all know the statistics. In nearly all Western countries between one-third and one-half of all marriages now end in divorce. At least one-third of all children now live outside a traditional family structure, either because their parents have divorced or because their parents never married in the first place. Some of these live in stable situations with two loving, though nontraditional, parents, but many more do not. More commonly, they live with one or the other of their natural parents. A great many of these children, the psychologists and sociologists tell us, will grow up damaged. Those who are products of divorce will suffer from depression, guilt, and a sense of inadequacy. They will do less well at school, fewer of them will go on to higher education, they will settle into unemployment or lower-status jobs than their parents, they will be more likely to commit crimes or take drugs, and they will themselves eventually make unstable marriages.[1] The adults themselves who have left the children in this state will be more prone to illness and depression and will, on average, die younger than if they had entered or remained within the married state (or within a stable relationship).[2,3]

It is of course true that some people are unsuited to marriage. A very few are more naturally comfortable with, and more likely to thrive within, a more solitary life. Others are lesbians or gays for whom what society means by marriage is not an option, though lasting relationships may be. It is equally true that *some* "one-parent families" are a success, providing safe and nurturing backgrounds for the children raised within them, though for the most part the statistics for these alternative arrangements are, if anything, more bleak than those for traditional families.

It still remains the case that the vast majority of us are heterosexual and wish to have lasting relationships. Most of us still wish to see these relationships sanctified by marriage and children. Without passing judgment on the alternatives, I want to concentrate on the kind of family the majority seek and try to understand why it is in trouble.

The traditional family as we have known it is in crisis. But even the grim statistics I have quoted reveal only a small part of the wider story of how this crisis affects us all. The statistics don't reveal that my own children feel threatened and insecure because of trouble in their friends' families. They don't reveal the amount of pain, misunderstanding, and frustration within a great many families that do stay together. They don't reveal the damaging effects on wider society and its stability through the breakdown of the most primary social unit. And they don't reveal how each of us is affected, directly, in our friendships and daily relationships and in our sense of the meaning and lasting quality of our own bonds. As American psychologist Judith Wallerstein has summed it up,

> Divorce has ripple effects that touch not just the family involved, but our entire society. . . . "Each divorce is the death of a small civilization." When one family divorces, that divorce affects relatives, friends, neighbors, employers, teachers, clergy and scores of strangers. . . . Today, all relationships between men and women are profoundly influenced by the high incidence of divorce. . . . Radical changes in family life affect all families, homes, parents, children, courtships, and marriages, silently altering the social fabric of the entire society.[4]

I think the official divorce statistics tell only a small part of the story because divorce itself is only the end result of a still deeper crisis within the family. Those of us who have not yet, or even those of us who never will, get divorced have had to face the underlying causes of this crisis, and the strains that ensue, within our own families. We have had to alter age-old assumptions and age-old habits and to some extent, if we have remained successful families, we have had to *reinvent* those families of which we are a part. It is that process of reinvention that I want to discuss here. It is a process necessary, I believe, to transform almost every "traditional" family into a family that can cope, indeed into a family that can *thrive,* in the conditions of postmodernity.

There are two different but inextricably related levels of family life that call for reinvention. There is, on the one hand, as the authors of *The Good Society* have seen, the whole question of the *meaning* of the family, the question of the moral—and I would add, spiritual—foundations of the various relationships within the family. These involve the meaning and moral and spiritual foundations of the intimate relationship between the couple who are at the heart of the family, and the meaning and moral and spiritual foundations of that couple's relationship to their children. In neither do the old foundations remain unquestionably sound. These foundations may have to be reinvented.

Second, there is the related question of the family's structure, of how the daily matters of cooperation, commitment, power, and authority are organized or enacted. These structural issues cannot be separated from the wider issue of meaning, but they are different. They are about how family members will relate to each other and negotiate their differences over opportunities or problems that arise day to day. They are about roles and responsibilities within the family, about the actual *construction* of the family, and about how the family relates to wider society. For most families, this structure, too, must be reinvented. Indeed, I believe it may have to be reinvented almost on an individual-case basis, family by family.

Traditionally, the family derived its meaning from several complementary sources. Until the Industrial Revolution, parents and children were very often an economic unit, sharing the responsibilities and the gain from common labor in the fields, or in some small craft or service industry, or in running a small shop. These family units were crucial to survival and were important elements in the larger economic structure of a town or region.

At the same time, well into the twentieth century, because there was little opportunity to control fertility, families were large and the nuclear family itself was caught up in a whole nexus of wider, extended relationships. Its meaning was derived partly from that nexus of relationship and from the fact that most of these relatives lived within a close distance of each other, knew each other intimately, and bore some responsibility for each other. These extended families themselves were embedded within traditional community structures. There was little of today's opportunity for casual or anonymous sex, little chance to escape responsibility for the products of that sex, and little scope for caring for those products (children) outside the family structure. All this gave the family a kind of natural, "automatic" meaning derived from necessity and wider relationship.

The family has also always had a biological meaning. The sexual relationship between husband and wife provided a steady and safe outlet for instinct, it gave the couple pleasure, mutual support, and a source of entertainment, and it provided the "pair-bonding" necessary to provide the couple's children with the long-term home and care required by the human young.

But beyond biological or economic necessity and social relationships, the family also derived moral and spiritual meaning from its religious foundations. Marriage between a man and a woman was considered a sacred act sanctioned, encouraged, and in different ways viewed as pleasing to God by every major religion. The bond between the man and the woman was seen as a binding covenant that expressed the holy nature of their union, the lifelong companionship and mutual care it would bring to them, and its expression, where possible, in the procreation and nurturing of children.

According to the Jewish mystics (as also in Hinduism, Tantric Buddhism, and Taoism), the sexual bond between man and wife had further, deeper meaning. It had the added dimension of being necessary to the completion of the Divine Nature (or Being) itself. It was through their union that God's own masculine nature (the Godhead) could unite with His feminine aspect (the *shekinah*). Thus for religious Jews, the sexual union of man and wife on the Sabbath is still seen as a holy duty (a *mitzvah*). The pleasure they give to each other is said to give Divine pleasure, and if the wife should conceive during this act God fulfills His part of the covenant by giving the child a holy soul.[5]

Christian marriage has never had quite the same religious significance. The celibate state has traditionally been held to be more holy than the married state, and Christianity has always been ill at ease with sex. The Christian God has no feminine aspect, so the dimension of Divine need is missing in the Christian union of man and wife. But the Church does nonetheless view marriage as a sacrament through which the couple acquires Grace, and the procreation of children is regarded as a holy act. The couple's union is held to be inviolable, it is meant to play a central role in the functioning of the Christian community, and, as in Judaism, infidelity is seen as a desecration—of both the social and the sacred bond that unites the couple. "What God has joined together let no man put asunder."

Nearly all these traditional meanings of family life and sexual intimacy have been undermined in modern and postmodern society. Very few Western families still function as an economic unit. On the contrary, in many Western countries having a family can put a couple at economic *disadvantage*. Where women were once confined to the home with many children and financially dependent on their husbands, most are now free to be independent if they wish. The whole value of staying at home and looking after children is now, rightly or wrongly, diminished.

A combination of birth control and geographic mobility has diminished both the size and the significance of the extended family. Far fewer families any longer find themselves woven into any kind of traditional community structure. Today the two-generation nuclear family can easily find itself far from relatives, rooted in the community only through the children's schools, and with few intimate friends to whom emotional or financial support for the family is an issue. The family no longer exists at the center of a whole nexus of social and blood relationships. It no longer derives its sense of meaning or identity from those relationships. It no longer extends naturally into the community.

Many people still long to have children, but their procreation is no longer the necessary by-product of sexual relations. With the world's population now reaching crisis proportions, they are not even a necessarily *desirable* by-product. Children are now a matter of *choice*. Their inevitable existence as the natural end of normal relations between a man and a woman no longer provides an automatic reason for embedding sex within a lasting marriage.

With so much of the necessity and the traditional, "automatic" rootedness of family life removed, the importance of a broad spiritual foundation becomes paramount. I say broad because the spiritual aspects of a couple's relationship to each other and to their children includes much more than belief in the given tenets of any organized religion. Our spiritual attitudes derive ultimately from our relationship to life itself, to its meaning and purpose. Our spiritual attitude to the relationships we enter, particularly sexual ones or those with people who are dependent on us, follows from our whole broad attitude toward identity and commitment, from what is meaningful and precious to us, and from what we feel ourselves to be responsible for. It follows from our sense of the sacred, from our ability to recognize that some things and relationships are sacred, and from our further ability to articulate, if only in faltering half-expression, from whence that sacred dimension comes.

In all these ways, I fear, both sexual and wider family relationships are currently in a state of spiritual crisis. The individuals who enter these relationships are themselves spiritually all at sea. I personally believe that this spiritual crisis is more fundamental and more threatening to family life (and to the individuals who might benefit from family life) than any of the more concrete changes that have come about in economic and social circumstances. I also feel that if we could get this right, if we could rediscover a meaningful spiritual foundation for the sexual and emotional intimacy between man and woman and for the commitment that binds them to each other and to their children, the family would reemerge—albeit, quite likely, in a structurally altered form.

A great many couples still have a nascent sense of the sacred about their wish to make the supposedly lifelong commitment of marriage. That is why, despite the stark decline in church attendance, so many still choose to have weddings performed or blessed in a religious ceremony. But these church (or synagogue) marriages are no more likely to last than their secular "town hall" counterparts. Indeed, it is increasingly likely the marriage will have been officiated by a divorced priest, minister, or rabbi.

We still turn to the established religions when we feel an occasion calls for special ceremony, but the ceremonies no longer transform us in a lasting way. In part (in very large part, I sometimes think) this is because the old religions speak a language to which we can no longer respond. But it is also true that *we* have changed. Our capacity for response to certain kinds of commitment has been muted.

Though we still have the impulse to make sacred commitments, the impulse to dedicate our lives to something or someone beyond the narrow confines of immediacy and the self, we find little supportive basis for such an impulse in the values or the philosophy of our culture. The philosophy of liberal individualism, derived in large part from the mechanists' atomistic picture of physical reality, stresses our separateness, our autonomy, our independence from all group or institutional layers of identity. As I noted earlier, it follows from this philosophy that "I" am not my role as wife or mother, "I" am not defined through my relationship to my husband or my children. "I," my self, is prior to all these relationships and roles. I am someone who has a great many rights but precious few duties except, of course, an almost sacred duty to nurture and coddle myself, a duty to "explore" myself and to "grow."

I may have the impulse to make commitments, but nothing in my individualist concept of self tells me that it is sacred to honor them. Indeed, as sociologist Peter Berger has argued, the whole concept of honor is "obsolescent." "Honour," he points out, "occupies about the same place in contemporary usage as chastity. An individual asserting it hardly invites admiration, and one who claims to have lost it is an object of amusement rather than sympathy. . . . At best honour and chastity are seen as ideological leftovers in the consciousness of obsolete classes, such as military officers or ethnic grandmothers."[6]

Honor, as Berger tells us, calls up a whole notion of the self as embedded within and defined through its relationships and institutional roles. But the individualist self is not so embedded. It is "free." It can at best be "constrained" or "hidden" by any symbols of or commitments to group or institutional identity or meaning. To return to my image of the dance, as individualists we are all soloists, all prima donnas. Any identification with the dance itself or with the other dancers might constrain our freedom of movement, might inhibit our free expression. Our solo movements have "dignity," but with no sense of belonging to the company they are without honor.[7]

This individualist ethic of course permeates our attitude toward sexual and family commitment. We have all heard countless young people say things like, "I will only stay with this relationship so long as I can get something out of it," or, "I will quit this relationship the minute it starts to go stale." Unfortunately the same attitude carries over into marriage, whether or not there are children. "I would never stay in a marriage that wasn't going somewhere for me," "I expect my marriage to fulfill me." In consequence, as a French jurist is quoted in *The Good Society* as saying, "Instead of the individual 'belonging' to the family, it is the family which is coming to be of service to the individual."[8]

We choose to marry because we think it may be good for us. It may give us what we need. All too often we choose to have children because we think they will be interesting, will provide us with meaning, will complete us, or will be "fun" to have around. The needy and searching self ("Modern man, almost inevitably it seems, is ever in search of himself,"[9]) gathers to itself whatever or whoever will satisfy—at once, if possible—an almost insatiable desire for personal fulfillment. This desire itself has been fueled not just by the individualist ethic in liberal political philosophy, but also by the radically individualist tone of modern psychology.

The whole ethic of the person given to us in Freudian psychology supports the view of the self as isolated. There is for Freud, as we saw earlier, just the individual and his "object relations." More than that, because of Freud's commitment to the sexual nature of all neurosis and the dominance of what he called the "pleasure principle," these objects that others are, are very often objects for my own gratification. I am bound by instinct and an urge to feel good, and if this instinct is repressed, I can become ill. Feeling good—feeling good *now*—takes on a terrible urgency.

*Ya, it's called mortal sin.*

Freud himself saw that if we gave way entirely to this sense of urgency civilization would be destroyed. Many of his followers tried to stress the more social nature of human beings and to encourage more responsible personal and community attitudes. But the dominant influence of all psychotherapies has been a narcissistic overattention to self and self-satisfaction. "All psychotherapies," says American psychiatrist Jerome Frank, "despite their diversity, share a value system which accords primacy to individual self-fulfillment or self-actualization. The individual is seen as the center of his moral universe, and concern for others is believed to follow from his own self-realization. . . . Our psychotherapeutic literature has contained precious little on the redemptive power of suffering, acceptance of one's lot in life, filial piety, adherence to tradition, self-restraint and moderation."[10]

All this concern with self, self-fulfillment, and self-satisfaction cannot but influence our attitude to others, to our sexual relations, and to our family commitments and responsibilities. As one critical psychotherapist has noted, we all too often view the quest for self-development in self-centered terms. We suppose that self-fulfillment does not come from sex, marriage, and relationship but instead is something that "seems to happen to individuals as they work on their own souls in the stillness of their private rooms."[11] Robert Bellah and his team add, "The obvious result of the growing emphasis on the primacy of personal satisfaction in family life is the rising rate of divorce."[12] When our only or our primary goal is personal satisfaction we cannot honestly commit ourselves to those very people with whom we might have found it. We are like a dog chasing its own tail. We are blinded to the emptiness of the chase by the sheer fury of our pursuit.

The individualist self can never find anything sacred in relationship because for him or her relationship is only a means to an end. The relationship itself has no value, no further reality of its own. The words of the marriage vow do not impinge, in part, simply because we do not have available the set of personal categories to understand them. How can we make sense of being "one in Christ" or "united in the People Israel" when, in the words of the ancient atomists, "It is silly to think that two things can ever become one"?[13] It is this whole failure of understanding that

I couldn't have done what I did, and still do, and maintain a marriage and raise a family! No way!

undermines both our sense of the sacred and any sense we might have of the sanctity of our sexual bonds and our families.

But individualism itself, the entire philosophical and psychological idea that human beings are separate from one another, that meaning can be private and that fulfillment can ever be purely "personal," is a misguided and misguiding illusion. It is founded, as we have seen, on a misguided notion of how physical reality itself is structured and on an ignorance of the creative dynamics of the physical universe. The atomists got it wrong. Physical reality does not consist of isolated parts. And not only is it *possible* that two things can become one, but such unity in difference is both the rule and the guiding principle of the whole of physical and biological evolution.

The quantum universe, as we have seen, is an entangled universe. Its many parts are interwoven, their boundaries and their identities overlap, and through their doing so new reality is created. As we saw in Chapter 2, when quantum system *A* unites with quantum system *B* to form quantum system *C, C* is *greater* than the sum of its parts. Something new, some further reality comes into existence through the relationship of the parts. Quantum relationship is not a means to an end. It is an end in itself. It is through relationship that the vacuum's (underlying reality's) manifold potential unfolds. In religious terms, it is through relationship that God or Being is realized.

Quantum reality gives us a powerful new model for the meaning and sacred dimension of our own intimate relationships. But as I have argued throughout, I think there is more here than just a good model. There is now evidence, as we have seen, that we really are quantum systems ourselves—our bodies certainly are, and it is very likely that our consciousness is as well. We too, indeed we more so than simpler living things because of the complex nature of our consciousness, have the capacity to create or to evoke reality through the relationships that we make. When I enter into an intimate relationship, my own being becomes entangled (interwoven) with that of the other to become *us.* This is very obviously true at the biological level when my sexual intimacy leads to the procreation of children, but it is also true at the psychological level.

Through sexual and emotional intimacy a couple create a further fact, the relationship that there is between them—their "dance." Their relationship, as in the metaphor of the dance and its dancers, evokes aspects of their individual potentials that were never before realized. It draws out qualities they never knew they had, indeed that they would not have had but for their relationship. But it also evokes potentiality from the underlying quantum vacuum, potentiality from the underlying ground state of reality—it evokes reality itself. This is a reality that can only be expressed through the existence of the couple as individuals and through their relationship. In religious terms, it evokes an aspect of God.

Our most intimate relationships are one of our sacred centers. Through the further reality that they evoke they offer one way that we can participate in the process of physical and biological evolution. There are other ways, of course. The creative artist, the writer, someone who follows a vocation or any creative activity with passion and commitment also evoke reality, as to some extent do we all simply by being conscious and "growing" our minds. But for most of us, our intimate marital relationships—the commitment required to sustain them and the potential latent within them—are our most powerful and most accessible route to the sacred. They embed us in creativity and meaning and in turn derive their own meaning from this. This, in quantum terms, is the first and most primary layer of meaning within the many layers of meaning that define family life. It is the layer associated with our sexuality, with the psychological and the sacred potential of our sexual being.

Some psychotherapists have described the committed sexual relationship at the heart of marriage as "nature's form of psychotherapy."[14] This is because it has a transformative power (and hence meaning) that I believe can only be understood fully in terms of the quantum nature of the person and quantum potentiality.

We saw earlier that people are often attracted to certain vocations because they need to confront and transcend their own problems, for example people with personal problems often becoming psychotherapists. The same is very often true in selecting a sexual mate—we are drawn to those who might complete us, who might help us to grow. This dynamic lies at the heart of Plato's

This reminds me like something from the Catholic Marriage Book of Common Doctrine

What about those of us who are monogamous and ask how and about gm a damn?

"Deanonimation" Doesn't everyone always "functionate"?

I've known some who were pretty fucked up and fucking up their kids

*I knew a psychologist-psychiatrist and professor at Cal Berkeley who really fucked up his own kids masterfully.*

myth of seeking the "other half." "And so, gentlemen, we are all like pieces of the coins that children break in half for keep-sakes—making two out of one, like the flatfish—and each of us is forever seeking the half that will tally with himself."[15] But Plato's view, as it is expressed here, sounds almost like the naive Hollywood notion that we fall in love and live happily ever after. The dialogue view of a committed relationship suggests, by contrast, that meeting one's other half is only the *beginning* of a lifelong process of discovery. It is a discovery of both self and other and of the relationship between them. This view is the heart of Martin Buber's description of *I-Thou* relationships.[16]

Thus, in any intimate relationship between two normal, imperfect people, after an initial "honeymoon" period each finds that the other raises deep problems that are often painful and that neither knows how to handle. Sex is the natural force through which these "unconscious," involuntary (uncontrolled) things rise to the surface. Sex has the potential to take us beyond what Freud called the "ego" level, beyond the calculated, the controlled, the polite, and the rational. It exposes our vulnerabilities, our uncertainties, and our points of growth. It takes us beyond our carefully constructed selves and mannered relationships to our (infinitely) *potential* selves and to the source of our creative dialogue with the vacuum that our relationships can be. But in consequence it takes us to a "far country," to a place where we don't know for a time who we are or how to respond. This can be both heady and terrifying.

At this point, unsupported by any philosophy or psychology that makes sense of or underlines the importance of the creative potential latent in the situation, many individuals or couples draw back. Many people learn to "control" sex, to remain detached or manipulative—those who engage in casual sex, prostitutes, the "Don Juans." Even couples who have committed themselves to marriage sometimes quit, walk out, or run away. Sometimes they continue, but estranged in an uneasy alliance that agrees to avoid problem areas. "We won't talk about that anymore," "We won't do that again." Very often one or the other may try to impose his or her needs or attitudes on the other, taking back "control" of the situation so that the other's presence no longer raises painful or seemingly insoluble problems. (These same defensive dynamics occur in psychotherapy.) This leads to confrontation and struggle, possibly unending or possibly resulting in the domination of one and the enslavement or repression of the other. The great unhappiness of one or the mutual boredom of both is the common consequence of these avoidance techniques. Either may lead to infidelity and possibly to the eventual breakdown of the marriage.

But faced with their own and the situation's creative potential, a couple *can* respond on a "dialogue" basis. They can *listen* to one another, accept both self and other without asserting either, simply be together and let what will unfold, trust each other and the energy flowing between them. If they do this, then just as in the dialogue process between any two people with different points of view, different priorities, different needs, and mutual but different senses of pain or urgency, something "magical" can happen. The new reality latent in the situation can emerge to change both the individuals and their relationship. Just like the photons "introduced" in the quantum laboratory which afterward get into nonlocal synchrony, the couple will change in synchrony (often through an intermediate series of role reversals). This is their "quantum dance." But there is a necessary presupposition to this process, just as there is in any dialogue process.

The couple must be *committed* to the process. Each must be willing to suspend cherished attitudes and beliefs and apparent lifelong needs. Each must be willing to "let go," to see through bad patches and not to respond or react too hastily. This requires a letting go of self, a letting go of expectations, and above all a commitment to the underlying potential of the relationship—ultimately a commitment to evolution itself and to the underlying potential of reality (the vacuum). This, as I have said, is in the nature of a spiritual commitment. This process may at times seem too stormy or confusing for a couple to handle alone. If so, one or both may approach issues more gently, or they may resort to the time-honored device of talking things through with a nonjudgmental friend (or a counselor). But always they must have the intention of returning to a direct dialogue with each other at some point.

There is no longer a list than the length of "bullshit terms" in psychology. Try withstanding a conference conversation!

See the Rule of St. Benedict. Some thing.

Though difficult to see, this is also, in subtle irony and paradox, a defense of celibacy for the dedicated priesthood.

## The Quantum Family

There is a reason why I have made what may seem a long digression into the nature and meaning of the sexual bond between man and wife. I feel certain that family life as a whole, and ultimately, the role and meaning of the family in society, follows directly from the kind of intimate relationship enjoyed by the couple who are at the heart of any family. Their relationship sets the pattern for all other relationships within the family. Its "style" will resonate through what family therapist E. Bruce Taub-Bynum calls "the family unconscious."[17] If it goes badly wrong, as we all know, it has the potential to destroy the family.

If a couple fail to develop the potential for growth latent within their *sexual* relationship, if they avoid or repress all the difficulties and challenges that arise, they may never get to that wider and more diverse level of creative potential latent within themselves. They, and their relationship, remain at the controlled (and even the contrived) level of the ego, the level of the known, of the actual. Their relationship is like a rigidly choreographed dance, its pattern and movements all laid down through plan or convention, its script written in advance and everything unfolding according to routine or expectation. Nothing surprising happens. This kind of routinized sexual relationship is very like all the relationships that we find within the homogeneous small group—tradition bound, reliable, sustained through repetition and familiarity, intolerant of much challenge or difference. The individuals who participate in it are cast in hard-and-fast roles. They harbor hard-and-fast expectations.

The kind of family that emerges from such a routinized sexual basis will be a family with the small group mentality and structure. The identity of each member will be rather rigidly defined through role within the family (wife-mother, husband-father, child), the roles will be arranged hierarchically, the family will rely heavily on tradition, repetition, and habit and will be slow to tolerate much difference from within or without. Relations within the family will be "polite," so as not too greatly to expose any member's weaknesses or vulnerabilities (or passions). There will be little scope for experiment, exploration, or eccentricity. The family will, in short, be a "traditional family."

The traditional family was supported by and embedded within traditional society. Indeed no other kind of family could have been the "basic social unit" of traditional society. When we describe this kind of family from today's perspective, we are at risk of demeaning it. Any social institution that works—and the traditional family worked spectacularly well until very recently—answers to the needs and expectations of the people living within it at the time.

Traditional society itself mirrored the characteristics of the traditional family and would have been ill served by any other. Any way of living will work if we commit ourselves to it and invest enough of our energy in it. We do make this kind of commitment more readily if we aren't seduced by other possibilities. A teenage child growing up in the 1940s or even the early 1950s would never have conceived challenging parental roles or authority in the way that today's children do. Few wives and mothers of that era ever thought to ask whether they should find some fulfillment outside the home, and society wouldn't have known what to do with them if they had. Similarly, few married couples thought to analyze the foundations of their bond because it, and the lifelong commitment it entailed, were social *givens*. But today there are few givens, few things that are not challenged or open to question. In postmodern society we find ourselves flooded with options. In this more complex and rapidly changing context, traditional families are as outdated as the societies that spawned and relied upon them.

The needs and demands of today's society are radically different—as we have seen they are the needs and the demands of the heterogeneous large group. Today's society is fast moving, complex, and plural. It calls for a response that is flexible and "ready for anything." It calls upon us to be quick thinking and adaptive. It has little time for tradition (perhaps sadly) and no taste for hierarchy. It is true that couples and families still need rituals and traditions, but it is important that they *evolve* or reinvent these for themselves, making them their own. And they must be prepared to alter or adapt them if circumstances change. The violence and the obsession with sex that increasingly dominate society's popular culture are at least in part, I believe, an unintegrated, undigested (Freud would say "unsublimated") drive to get beyond the "stuffy" ego-dominated relations and routinized feelings of traditional society and traditional, hand-me-down rituals. They are in part a cry for something more "raw," something closer to our creative potential. The whole of postmodern society is struggling to reinvent itself.

All these changes in society impinge directly upon the family. They come into the home through television, through the children's schools and friends and youth culture, through our experiences at work and relationships with friends. They permeate the whole culture in which the family is situated—a culture that is hell-bent on tearing down all the idols and the institutions of traditional society and its traditional authority. They also, as I have suggested earlier, come from the whole weight of modern psychology, from its attitude toward sex and relationships.

Freudian psychology has not helped us to see any way in which a more creative sexuality could be the basis for a renewed family life. I am not speaking of variety learned from sex manuals or exotic positions, but of a willingness to bring our actual feelings about the other into the sexual relationship, a kind of emotional sincerity and freshness that is risky but creative. Freud himself saw that the sex drive could liberate us from the stuffy constraints of traditional society (the Ego), but he feared the consequences. He had no solid model for the sublimation process that he believed could convert a liberated sexuality into a creative force. His pupil Reich threw off all inhibition and argued that free and unconstrained sexuality was utterly essential to psychic health. Reich believed that human beings are naturally polygamous, that few "permanent" sexual relationships have the staying power to last more than four years, and that to try to extend them would lead to neurosis.[18] In consequence, he argued that marriage as a social and legal institution has little to do with the individual's quest for sexual fulfillment.[19]

The Freudian legacy for the family has on the whole been to encourage, implicitly if not always so explicitly as Reich, the view that marital fidelity is a sacrifice or an asceticism, a price that we pay to keep our families and our society together rather than a positive and creative end in itself. It is seen as the enemy of diversity, a sometimes unbearable or boring counterweight to our urge for exploration and growth.

In the face of all these wider social and cultural upheavals, it is impossible for the family to carry on in the traditional mold. A few—the Amish, Hasidic Jews, and Muslim fundamentalists—try to avoid the problem by isolating themselves. They cut themselves off from the outside forces that threaten their way of life. If we don't favor isolation, we can at best cling to such families (and to the routinized, habit-bound, and role-related sexual "intimacy" on which they are founded) with a nostalgia similar to that which the Greens feel for traditional small-village life. All of us may sometimes feel the need for this being cocooned when we are insecure or overstressed. We can't "grow" all the time. But if this is the only, or the dominant, attitude within the home, we will look "back" on the family as a bucolic village from which we came and to which one day we dream of returning, a safe haven for when the storms of life become too demanding. If so we may be fated to rebel against it. We will take our adventures and do our growing elsewhere and with others. Such a fossilized family, should it survive intact at all, can play no creative role in the evolution of society, nor in the evolution of ourselves.

I believe that postmodern society *needs* the family, but it needs a radically new style of family life to be the building block and the training ground of the new society we are creating. It needs a family that is founded on the kind of committed intimate relationship that puts its participants in touch with their potential selves, with their areas of possible growth. As we have seen, this kind of relationship (where we not only seek but are willing to meet our "other half") calls upon those involved to confront difficult issues, to face pain and uncertainty and the unknown. It is not "comfortable" or "safe," but it can be creative. It releases the potential for growth latent within us all. D. H. Lawrence portrayed the gradual shift from the traditional to the postmodern style of relationship, with its extra possibilities and extra pitfalls, in his two greatest novels, *The Rainbow* and *Women in Love*. He made some

*Trying to better this "The White House"*

*His wife was a masculine horse!*

effort to live by his convictions in his own stormy but creative marriage.

A quantum understanding of human consciousness and the person helps us to see that there is an infinity of potentiality "sleeping" inside us all. This potentiality is evoked through our experiences and our relationships. There is no end to the persons that each one of us can be. Most of us feel this about ourselves, but we often have difficulty realizing it about others. Too often we take others at face value, as the given. We cast them in roles and relate to them as though they had no further dimension. When we thus trap ourselves in a habit-bound, routinized sexual relationship, we sometimes feel that we must "go further afield" if we want to evoke some new side of ourselves through relationship. Reich thought this was inevitable—hence his argument for polygamy. But the quantum nature of the person and the many-layered quantum potential latent within our relationships gives the lie to Reich's view.

I have, from time to time during my seventeen years of marriage, been attracted to some other man. Each attraction has promised to evoke some side of me that I didn't know existed and the heady experience of this has in turn evoked fantasies of abandoning all and running off in pursuit of my passion. This usually happens at times when the relationship with my husband has come to feel routine, perhaps even boring. I have always told my husband about these attractions. The conversations that follow are uncomfortable for us both. They raise pain and the specter of unfulfilled expectations on both sides. But my husband always *listens,* and through his listening something happens between us. Somehow I come to see that my husband himself possesses the quality I was attracted to in another man—though neither he nor I was aware he possessed it until our conversation. Indeed, the conversation itself, the honesty of the dialogue, evokes a potential latent all along and this new quality enters *our* relationship. The temptation to infidelity dissipates at once.

The temptation to infidelity is a temptation to grow and to explore the self. It is a need to feel that unrealized potentialities within the self can be evoked. Such exploration of self through diverse others is probably necessary for young people who are just feeling their way in growth and relationship. We may all go through such a period when a few trial (but honest) relationships help us to mature. But our greater maturity brings with it the potential for a deeper growth possible only in one, lifelong, committed relationship.

In our one "quantum" relationship we can have many relationships. Through an honest, and at times a necessarily risky confrontation of our needs and feelings, we can find the "man" (the man with all potentialities) within our husband or the "woman" within our wife. We can find what Jung would call the "archetype" of the masculine or the feminine within our beloved.

Through one lifelong committed relationship of this sort we can better explore the many potentialities within ourselves than through any number of single-faceted affairs. In the detached attitude that usually accompanies casual sex, our problems, vulnerabilities, and conflicts usually get left aside. In uncommitted affairs, we usually quit just at the point where (and usually because) they begin to arise. Indeed, many people quit at exactly the same point in one relationship (sometimes even in one marriage) after another, destined always to repeat the same incomplete and ultimately unfulfilling drama. If, on the other hand, we stay the course in one committed, lifelong relationship, and do so creatively, we must confront our fears, our weaknesses, and our insecurities. These become growing points rather than something from which we run.

The route to potentiality within a relationship need not be through temptation. It may lie simply in confronting pain or boredom when it occurs. (Boredom itself is usually a symptom of blocked energy, a symptom that we have got stuck.) It may lie in a commitment to avoiding routine and habit—going new places together, doing new things, sharing new interests. But in all these cases, the kind of relationship that it evokes can be the pattern for a whole wider family structure that is itself creative.

Where the traditional family, with its heavy reliance on roles, hierarchy, habit, and polite, routinized expression of feelings, is like a rigidly choreographed dance, the "quantum family" is more like the benzene ring I described in an earlier chapter. In the benzene ring, we can recall, several carbon atoms and several hydrogen atoms share their electron rings (i.e., share their identities) in a constantly resonating, shifting, and changing set of *possible* configurations. Each atom relates through its *potentiality* (its wave nature), so the whole ring exhibits more flexibility than any classical fixed role structure could do. An equivalent family structure would have the poise always to be *somewhat* (not entirely!) in flux, living through its potential and thus creating itself and its members as it freely explores its experience through dialogue. (It is worth recalling that the benzene ring, with its superposition of possible resonance states, is more stable than any one of its "actual" configurations would be on its own.)

How each family reinvents itself if it relates in a nonroutinized way in dialogue with its experience and between its members will vary from family to family. There will be no "ideal" structure or "ideal" role fulfillment, only the frequent presence of dialogue and the realities it can evoke. I can think of one very big example from our own family's experience that may help to make this more concrete. It upset us all at the time, it still leaves us not knowing quite where we are, and yet it has both defined our family's style and strengthened the various relationships within it.

Until a few years ago, since the birth of our children, we had been attending a local Christian church fairly regularly. We knew the vicar well, we were a part of the church community, and the children played an active role in the service. Then one day, when they were four and six years old, both children suddenly announced that they wanted to be Jewish. Each had a best friend who was Jewish and this had strongly influenced them. At first we laughed and said, "This family is Christian."

As the children persisted with what became a campaign, our own arguments became more determined. We told them that we had too much "invested" in the local church community simply to pull up our roots and that when they were sixteen they could join what religion they liked. We continued insisting that they attend church on Sunday mornings, but they sat in the pew with their arms folded and their eyes rolled toward the ceiling and kept repeating, "This is ridiculous. Jews don't go to *church*." We were in a standard confrontation situation.

Because the children were so young, we could have gone on

forcing them. In time, they would probably have ceased their campaign. But something made us feel this could damage them. We decided to listen. We let them put forth their feelings and their point of view, and we promised that we would consider it. I promised them I would go see a rabbi (I had myself converted to Judaism in my early twenties, though since my marriage I had again begun attending Christian services with my husband) and ask him how we might do a few Jewish things at home. We also promised to take them to a synagogue. Though none of us could have foreseen what would happen, the events that followed changed the whole basis of our family religious life and have greatly influenced our work.

Though my husband had a lifelong commitment to Christianity, and I a very large commitment to Christian culture, we all began attending the synagogue and seeing a great deal of the rabbi. The children enrolled themselves in Hebrew school on Sunday mornings, which meant they could no longer attend church. For a while we continued to go without them. But slowly we began to rethink our own position. We reached beyond both Christianity and Judaism to something underlying both and came to feel that our true commitment was to that. In the words of the Dalai Lama,

> I consider the essence of all religions is a good heart, a peaceful mind, compassion, forgiveness, respect for other lives, a sense of brotherhood, sisterhood. These are the essential message of the various religions. We use the same material, like gold, but we can change the shape according to our practical needs. So similarly religion has the same essence, but the requirement in our daily life is of new approaches.[20]

All this is just as well, because after a recent trip to the Far East the children turned everything on its head again by announcing they wished to become Buddhists! We've got to a place where we don't know what to write on the school forms when they ask us to state "religion," but we all have a partially articulated sense that there is something we value and that it is sacred. It is that "gold" of which the Dalai Lama speaks, some underlying source that has the potential to take many forms. In the process, we now realize, we have reinvented not just our religion but also ourselves.

If a family can live at the level of its potential (that is, at subego level, "before" the level of habit and routine), reinventing its private "institutions" (its rituals, its ceremonies, its daily routines) as it goes along, it is likely to be a vital and creative family that nourishes its members' growth. A continual openness to dialogue can transform the gentle sexual energy (energy from beyond the ego) that permeates any close family (such families are likely to be what Bellah and his team describe as "warm" families—families that "pay attention"[21]) into a whole range of emergent realities. Some of these may appear "negative"—pains, conflicts, insecurities, petty or unworthy emotions. Others will be more obviously "positive"—ideas, "projects," energy, and enthusiasm for doing things. But all are in the end creative. As Martin Luther said, "Man never flies so high as when he knows not whither he is going."

But perhaps most important for the family's place in our post-modern society, such ongoing dialogue and exploration can leave the family "open" to experience, ready and able to integrate and to transform the constant barrage of diversity, pluralism, rapid change, and even chaos being thrown up by the culture at large. Such families, being "where it's at," centers of the storm rather than safe havens, can be the basic social units of the emerging society. They can be society's integrative center, a vital institution that digests the chaos.

The reinvented family must be open to experience, but it must also open out into the world. The nuclear family itself is a very private, close-knit, and intimate group—at least it has the potential to be so. Its own intimate internal relations do not impinge directly on society as a whole. For that there must be bridges, intermediate relationships or points of contact that "connect" the family to society. In traditional society this bridging role was played by the large, biological extended family. Such large extended families, all living within the same geographical area, seldom exist anymore in Western countries, as we have seen. Even if they did, I doubt they could provide the kind of bridge required today.

The traditional biological extended family, though large, had the dynamics typical of a homogeneous small group, with the small group mentality and way of relating. It relied heavily upon habit, tradition, and convention. It did not call upon its members to cope with much rapid change or diversity. But that is the main challenge facing both the family and society today. We have now, I believe, to reinvent that extended family, to surround ourselves with intimates who think and relate more in the manner of the large heterogeneous group. This new extended "family" must bring diversity into the home and carry the nuclear family's transforming creativity outward into society.

I believe we must consciously *construct* our reinvented extended family by "co-opting" members who are not necessarily relatives (though some may be) but chosen, rather, on grounds of shared ideals, shared interests, common need, or simple affection. We must share an important part of our lives with these co-opted members, see them often, share activities and rituals (birthdays, weddings, funerals, Sunday lunches, and holidays), and engage in dialogue with them. We must feel "free" with them, able and willing to express ourselves with little more restraint or inhibition (little more polite "control") than we would express with intimate family members. This is important if we are to relate to them at that creative level of potential rather than in a stiff, formalized way.

The more diverse the characters, interests, and backgrounds of these co-opted "family" members, the more diverse and stimulating will be the bridge they make between the nuclear family and wider society. They must be our closest community. When we commit ourselves to them and they to us (through rituals or bonds or promises made almost with the same spirit as the marriage vow), our "sphere of intimacy" is extended and our inner dialogue brought onto a wider stage.

As I write this, I think both of "the community of Christ" and "the People Israel." Surely something like this is the ideal in both these organized religions. It must, I believe, be an ideal toward which we strive in focusing on the role, the meaning, and the vitality of the family within a "quantum" society.

# 14. Toward Quantum Consensus: Unity in Diversity

*It may be impossible to recreate a moral consensus in an advanced, plural democracy. Nobody has ever done it and there are good reasons for thinking it cannot be done.*

Bryan Appleyard*

Politics, as we noted earlier, is usually defined as the art or process of conciliation between different interests in society. That is the accepted democratic definition. More pessimistically, politics is sometimes defined as the art of learning to fight one's own corner, the art of getting one's own way. That is the more manipulative or adversarial definition, a definition more associated, at best, with accompanying arts of persuasion and at worst with conniving, bullying, or even tyranny.

The conciliation sought by democratic politics, whether it be the small day-to-day politics found within the family or the small group, or the large-scale politics of national and international affairs, is arrived at ideally through some form of consensus. We will hope to agree on some course of action or statement of belief common to all or, at second best, we will agree that *some* of us will accept graciously to go along with the plans or beliefs of the majority. We will, as the English say, agree to disagree.

*The Independent, 4 February 1993.

But, as Bryan Appleyard has said, in our increasingly pluralistic societies we find the arrival at any form of meaningful consensus difficult, if not impossible. This is true not just at the level of politics but more fundamentally at the deeper levels of culture, morality, and spirit. We have different lifestyles, harbor different aspirations and values, worship different "gods." There seem to be ever more things about which we must agree to disagree, to the point that our "consensus" becomes an empty form. This, I have argued, has been exactly the fate of much liberal consensus—it leaves out (privatizes) so much that we care about that we are left sharing little more than a moral and spiritual vacuum.

There are no more ties that bind, no cement of society. This leads to fragmentation and, inevitably, to the kind of politics that descends to corner fighting, conniving, and manipulation. On the streets, many suspect, such personal corner fighting contributes to the rising crime figures that blight life in so many of our cities and small towns. Where there is no obvious common bond with the other and no common agreement about what is right and wrong, what is valuable and what is cheap, there is little common ground on which we can draw as a reason to respect the other's person or property. As a British theater director who works with convicted offenders comments,

> In a society less held together by shared moral absolutes than by order imposed from above, the individual has fewer scruples about breaking the law. The deterrent is not so much the likely guilt as fear of reprisals. And since everyone nowadays seems to be out for themselves, crime can be said to be justified, necessary even.[1]

We are left, then, with what seems an impossible conundrum. In any postmodern democratic society, pluralism seems inevitable. We will not, without enforcing massive conformity and assimilation, population movements or apartheid, ever again live in the kind of homogeneous societies that were once the norm. Yet it seems that democratic freedom thus breeds the source of its own fragmentation and must surrender its quest for the sort of consensus that gives shared life any meaning and democratic politics its highest goal. Pluralism is at once made possible by democracy and at the same time seems fated to undermine some of the best fruits of democracy. The *best* we can hope for is an array of loose compromises and a great deal of tolerance.

These, at least, are the sad conclusions to which we are forced if we have only the old mechanistic model that individuals or individual homogeneous groups are so many atoms bouncing about and colliding in some featureless social space. They are the conclusions to which Bryan Appleyard and many others writing on the state of contemporary society are drawn. If society consists only of selves and "others" or of "my group" and "your group," the emphasis being on what philosopher John Rawls calls the "distinction between persons" or Robert Nozick "the fact of our separate existences," then we are left with an unpalatable choice between two extremes. We can, like Thomas Hobbes, argue that some sort of consensus must be *imposed,* that the atoms must be forced to act in some kind of harmony. We can limit free speech and personal movement, hire more police, and thus give up our aspirations to democracy. Or we can, like liberal philosopher Stuart Hampshire, live with the fragmentation of pluralistic democratic society (and its crime statistics) and accept that "justice is strife."[2]

But there is a third way to look upon pluralism and difference and the search for consensus (moral and spiritual as well as political) in a pluralistic society. That is the way advocated by this book. *If the book could be said to have any one overall theme it has been exactly this—to explore how we might both celebrate our diversity and at the same time find some creative unity in our differences.*

We have seen that in a quantum society there are no "others," that any sharp distinction between self and other is a false and dangerous one. We have also seen that diversity is no threat to consensus. On the contrary, our differences, the qualities and beliefs and lifestyles that distinguish each of us as individuals or groups, have turned out to be the very stuff on which a deeper and more meaningful consensus can be built. They are the stuff out of which all evolution, personal and social as well as physical and biological, proceeds. The quantum promise of an underlying unity celebrated through our diversity has been the theme by way of which we have seen how we can redefine our communities, recapture the meaning of our freedom, transform our politics, and reinvent our families.

This theme has been explored by proposing a radically new model for what individuals are and how we relate in society. The model itself is drawn from the many rich images and metaphors suggested by quantum reality, and from the further possibility that human consciousness itself may have quantum origins. I think that reminding ourselves why *some* new model is necessary and how this quantum model lays new foundations for consensus may be the most useful way to sum up the book's main ideas and arguments while at the same time giving us a better understanding of pluralistic consensus itself.

*The Need for New Thinking.* Each of us in our personal lives calls upon images, analogies, and metaphors that help us to articulate our experience and to bind that experience into a coherent unity. These images are not themselves thinking, but they are the foundation on which thinking rests. We have, for example, passionately held and defended images of freedom, justice, love, integrity, honor, even scientific facts, and their opposites. Each image has a whole range of possible expressions and interpretations. From this vast mythic or dramatic seedbed we may extract more precise and limited concepts and categories in terms of which more "rational" thought is possible. Our original overall vision resting on these images evolves, becomes more detailed and more realistic. There are controversies and debates within the shared, broad framework. Thus visions are born and grow. The same process of paradigm birth and growth can be seen in science, philosophy, religion, and politics. In our social life, above all, we need some such shared vision if we hope ever to share a consensus.

But eventually a shared vision—a "paradigm" to use Thomas Kuhn's expression[3]—becomes exhausted. Any one vision or paradigm is finite, a bucket that can hold only some of the inexhaustible well of Truth. New questions present themselves that are insoluble in the old terms, that cannot even be formulated in the old terms. Discussion becomes clever but sterile wrangling among those who still accept and try to elaborate the old presuppositions.[4]

Each contestant can criticize his rivals' positions without being able to articulate why he is ill at ease even with his own. As political philosopher Michael Walzer says, once the images and analogies on which we have based our theories are called into question, it becomes impossible "to think the old thoughts." The old paradigm dies, becoming what Walzer calls a "conceit."[5]

The pre-Renaissance Western vision of the world was of a single, harmonious system created by God. Church and State were God's representatives on earth. Society was often described as a "body politic," an organic unity with the king as "head" and in which civil disorder was seen as a "disease."[6] But with the Protestant Reformation, the Copernican revolution in science, and the gradual loosening of static social structures, this vision became less compelling. It was replaced by Hobbes's mechanistic vision of people behaving like atoms in the new mechanistic physics, each moving about chaotically and clashing with others. Now order had to be imposed by force or negotiated by Social Contract. Mechanism and its accompanying individualism were the pillars of a new paradigm that held unchallenged for two hundred years.

Today we have individualistic and radically opposed communitarian models of society. Each can point to obvious defects in the other, but neither seems capable of fresh thinking about the problems raised by increasing pluralism. The materialist, modern paradigm has in this century begun to show the familiar signs of exhaustion. Politics itself is increasingly irrelevant to many people.

Certain presuppositions, of course, still remain powerful, for example the alleged split, originating with Descartes, between the private, subjective realm of individual experience and the public, objective realm of matter, and the accompanying notion that the only valid form of understanding is the "objective" observation of data according to the criteria of the old, Newtonian sciences. But these presuppositions have been less rewarding in psychology, sociology, and the arts, and most obviously as a model for understanding close relationships. These, coupled with the loss (for most of us) of resonant images from nature and religion that once gave meaning to our lives and work, have led to starvation of the contemporary imagination. As philosopher Susanne Langer expresses it, "The springs of European thought have run dry."[7]

There are many antirational attempts to counter this state of affairs, from nationalism and deconstructive postmodernism to the wilder New Age fads, but none seems capable of articulating a coherent and wide-ranging new vision from which we can derive a new set of meaningful and resonant images. Nor do any of these reflect the radically new kind of thought emerging within twentieth-century science. The images derived from this science, particularly from quantum physics, are perhaps the most potent and most fully articulated available, and reflect what may be the most profound achievement yet of recent Western culture.

*Recognition That We Are Both Self and Other.* Quantum reality, we have seen, has the potential to be both particlelike and wavelike. Particles are individuals, located and measurable in space and time. They are either here or there, now or then. Waves are "nonlocal," they are spread out across all of space and time, and their instantaneous effects are everywhere. Waves extend themselves in every direction at once, they overlap and combine with other waves to form new realities (new emergent wholes).

This "wave/particle dualism" offers us a powerful new model for seeing ourselves both as individuals, distinct and precious and effective in our own right, and at the same time as members of wider groups through which we acquire *further* identity and a wider capacity for creative relationship. If human consciousness is indeed quantum mechanical in its origins, then each of us literally may have a "particle aspect" and a "wave aspect," what philosopher Thomas Nagel calls our "personal" and our "impersonal" aspects.[8]

Through our personal aspects we cultivate what is unique about ourselves and our talents. We exercise our freedom and our responsibility. Through our impersonal aspects, our wave aspects, we overlap with others and find new parts of ourselves through our relationships with them. We literally *grow,* or construct ourselves, through our relationships. In some very meaningful sense we *are* our relationships. This is the quantum basis for what American political scientist Tracy Strong calls "the codetermination of self and society."[9] In moral terms, I am not my brother's keeper. I am my brother.

The same quantum logic applies to groups. There is not just

"my group" and "your group" but the emergent society, the "dance," in which all groups discover and explore and create themselves as they share their public space. That dance is the source of their common score, their consensus.

*Living at the Edge.* In quantum reality, things are determinate when they have been measured. Measured qualities are "pinned down," fixed in a given state. Once something has been measured, it no longer has an internal "freedom to become." It is now just as it is, an actuality rather than a potentiality. It lives out of one side of itself, it is *either/or.*

Before they are measured, many properties or qualities of a quantum "thing" are indeterminate. They are in every state at once. These indeterminate qualities exist in a state of spread-out possibilities and have an internal freedom to become fixed in any one of them. The quantum "thing" is in a state of *both/and.* Thus in the famous case of Schrödinger's cat, the cat is *both* alive *and* dead before we look at him, but once we open his box he is *either* alive *or* dead.

At the quantum level measured, or determinate, qualities relate *externally,* but they do not get into emergent correlations. Their internal characteristics do not change with the relationship. It is when quantum qualities are in the unmeasured, or indeterminate, state that they get into "sync." They become defined in terms of their relationship and the relationship gives rise to new realities.

Similarly, the quantum model of society suggests that we, too, have the capacity to be and to relate in both a determinate (external) and an indeterminate (internal) way. When we harbor fixed attitudes, allow ourselves to be or cast others in rigid roles or stereotypes, when we have fixed expectations, live our lives as a set of well-worn habits, follow routines or adhere to rigid bureaucratic rules, we are living out of our determinacy. We restrict ourselves to the level of here and now and actuality. This is sometimes necessary, perhaps even sometimes desirable (the trains must run on time!), but it is not creative. It fixes us and our relationships in place. If it becomes the rule, it exposes us to the risk of growing stale or bored. It denies us the opportunity to explore ourselves and to become creative members of an emergent group, family, or community.

By contrast, when we live "at the edge" (quantum self-organizing systems, we recall, are poised at the edge between order and chaos), when we accept the risk of our freedom and allow ourselves to be open to new experience, open in our attitudes, open to the many possibilities within ourselves and others, ready to reinvent ourselves, our relationships, and our families, we live out of our indeterminacy. We live at the level of our potentiality and remain fresh like children. Metaphorically, we live at the level of poetry rather than prose. Ambiguity and ambiguous (multifaceted, multilayered, suggestive) communication are our friends, not our enemies. We stand poised toward internal relationship, community, and an emerging consensus.

*Celebration of Diversity.* The Schrödinger wave equation describes unmeasured quantum reality as an infinite spread of possibilities, or potentials. Each quantum "thing" has the potentiality to go this way and that, to be here and there, now and then. Until it is measured, quantum reality lives out *all* these possibilities simultaneously. Each may contradict all the others and yet all are necessary to a full description of what and how a quantum "thing" is.

Thus the cat again——he is alive *and* he is dead and both descriptions are necessary to tell us about him. (Both are described in his Schrödinger equation.) Light is *both* a wave *and* a particle, though we can never see both at once. The Schrödinger equation is quantum reality's own fundamental celebration of diversity. Yet because it is a description of just one underlying "thing," the equation ultimately celebrates a unity in the diversity.

Similarly, as Gödel's Theorem states, the whole truth can never be caught in a finite net. There is always more to truth than can ever be expressed in any one statement or system of belief. According to the *Tao Te Ching,* "The way that can be expressed in words is not the eternal way." There is not one way but many ways and *all* are necessary to describe *the* way.

In society, the larger and richer the range of diversity the greater the opportunity for that society to express its own underlying potential. The greater and richer the range of my experience the closer I am to realizing my inexpressible true self.

Inside me, inside each one of us, there is an infinite range of potential selves waiting to be evoked through relationship to others. The other is my opportunity, my necessity for growth. The otherness of the other, his *difference,* is a possibility sleeping within myself. I need the Muslim to be a Muslim, the Christian to be a Christian, the Jew a Jew. I need to be me, to hold my values, and I need you to hold yours. On a quantum understanding of the phrase, agreeing to disagree is agreeing upon something very fundamental indeed. *That* is the agreement upon which we can build our pluralistic consensus.

*Commitment to Dialogue.* Diversity on its own can confuse and fragment as well as enrich. In both the physical and social worlds differences must meet and be integrated if they are to lead to new creative reality.

At the level of quantum reality, differences that are in a determinate, fixed state relate only externally. They have the potential to clash. But differences that are in an indeterminate, wavelike state meet, their wave fronts overlap and combine, and the potentialities carried by each wave give rise to a new reality that is greater than the sum of its parts.

Through the relationship of two quantum systems, some of whose qualities are in an indeterminate, potential state, those unfixed qualities are evoked into actuality. Thus the two photons in the nonlocality experiments each acquire a polarization *as a pair* that neither had on its own. In the metaphor of the dance, the dance itself has characteristics and the dancers a kind of moving, coordinated grace that no one dancer had on his or her own. Their relationship evokes the reality. In quantum terms, this is known as "relational holism."

In the brain something similar happens. When the brain first perceives a very heterogeneous field of which it cannot make sense with its habitual perceptual categories, it puts itself "on hold." All the diverse data are held together in the limbic system while the brain goes through a process of deconstruction leading to resynthesis. Deconstruction is very much like letting itself get into an indeterminate state——it lets go of its old concepts and categories, it "decides" to look at the data afresh. Then, during the process of resynthesis, new concepts and new categories are evolved that can integrate the diversity with which the brain has been challenged. New neural pathways are laid down, new perceptual skills evolved.

We have seen that in society both the dynamics of quantum relational holism and those by which the brain integrates new heterogeneous data are remarkably similar to the dynamics of a good dialogue. We come to dialogue with an attitude or a belief or a plan that we are willing to put "on hold." We are willing to put ourselves into an indeterminate state—to be open, to listen creatively to the other, and to look at the issues concerned afresh. We, in effect, let go of our familiar concepts and categories. The other does the same. And through the process of dialogue a resynthesis occurs. Each of us learns new concepts and new categories. Each of us arrives at a new understanding and perhaps even a new position that neither could have achieved on his own. We do not necessarily agree at the end of dialogue but we do, in the quantum sense already described, agree to disagree. Our willingness to engage in dialogue, the emerging pattern of the dialogue itself, is our common ground. The meeting of our differences is our consensus.

If dialogue is to be the heart of our pluralistic politics and our new consensus, we must create the conditions that nurture it. There must be times and places where differences can meet and get into dialogue. Some of these we have discussed at length. Some we can create as individuals or small groups. Some our politicians could and should create. Some are created by the existence of the arts. The ancient Athenians met at the theater as meaningfully as they met at the Parthenon. Theater was a crucial part of their ongoing dialogue and the unfolding reality of the Greek psyche and polis.

In today's society there is the potentiality for dialogue in a free press and through the media, if these are used creatively for this purpose. But public leisure facilities and public spaces where differences can meet and "listen" to each other are important, too. At play we are often more "indeterminate" than in our daily lives, more open to meeting others. Hence the old British belief in the importance of the playing fields of Eton as the place where Empire was spawned. In shared public spaces or communal public events where people meet, hear each other's voices, watch each other's movements, sample each other's ways, the conditions are created for an emerging common reality in which the differences flourish and enrich. All the movements add to a dance which is "our" city, "our" people, "our" nation, "our" world.

*Commitment to Our Common Ground.* Physicists can now describe an underlying, common reality that is the "ground state" (lowest energy state) of all that is in the universe. It is known as the quantum vacuum but it is not empty. Rather, the vacuum is replete with all potentiality. In the language of physics, everything that ex-ists, that "stands out" as an actuality, does so as an excitation of the vacuum. We have used the metaphor that the vacuum is a pond that contains all being, and everything that exists is a ripple on that pond.

We, too, our bodies, our minds, our relationships, our groups, our cultures and societies are, strictly speaking, excitations of the vacuum. We could not *be* otherwise. In religious terms, we suggested that the vacuum could be thought of as the God within, and all of us—all things that exist, all "ripples on the pond"—are then thoughts in the mind of that God. We are all "brothers and sisters" in the same source, stuff of the same substance. Our differences are the expressed potentialities of that common source. They are its unfolding reality. The vacuum is the ultimate unity underlying all those differences and the ultimate source of their meaning. If we become aware of this, if we commit ourselves to the task of reality creation, if we see ourselves as agents of the vacuum's (God's, or Being's) unfolding, we at the same time commit ourselves to the value and the meaning of the other's way, too, to the value and meaning of diversity.

*Commitment to the Future.* Evolution is the physical process through which all that is in the physical and natural worlds unfolds. All evolution proceeds through a process of using variation leading to selection leading to further variation. We have seen that there is now solid reason to suppose that our universe itself is a product of a still larger-scale cosmic evolutionary process and that this cosmic evolution proceeds in very much the same way that biological evolution proceeds. We have suggested that there is good

*Quantum Theology*

reason to suppose that consciousness, too, evolves according to the same pattern and indeed, that the quantum dynamics on which consciousness itself may rest are built into the unfolding process of evolution at whatever level it takes place. Thus there appears to be one universal principle of evolution at work at every level of reality, physical, biological, and social.

Evolution is our future. It is the process through which we get to our future, and by committing ourselves as fully as possible to that process we commit ourselves to the future. But commitment to evolution is a commitment to the value of diversity. Evolution feeds on diversity, it needs diversity. That need, a universal need, is the ultimate reason why each of us and the societies in which we live need "the other" and his "otherness." The consensus that we need is a consensus in which many voices are heard and in which those voices remain distinct, yet all are, if you like, voices in the same choir. To return to the central metaphor of this book, evolution needs its many dancers, but all are dancers in the same dance.

So I am a dancing Wave and Particle!

Well, when I die, this Particle will Wave Good-bye! And if I do come back, I don't care if I'm famous, but I do insist upon being talented, good-looking and rich!

Remember Donna Summer's song "The Last Dance" popular during the disco era? This is mine!

# • Notes •

---------------------------- **FOREWORD** ----------------------------

1. Bertrand Russell, 1957, p. 45.


--------- **CHAPTER 1 WHAT IS A "QUANTUM SOCIETY"?** ---------

1. David Bohm, 1990, p. 271.
2. Vernard Foley, 1976, p. xv.
3. Quoted in the Introduction to Adam Smith, 1986, p. 15.
4. Giorgio de Santillana, 1946, p. 249.
5. Vernard Foley, 1976, Chapter X.
6. Ibid., p. 22.
7. Vilfredo Pareto, 1935, Secs. 126–38; Sec. 2022.
8. Colin Blakemore, 1988.
9. Vernard Foley, 1976, p. 202.
10. Thomas Hobbes, 1983, Chapter XIII, p. 145.
11. Jon Elster, 1979.
12. Jane Mansbridge, 1990, p. 3.
13. Jacques Monod, 1972, p. 172.
14. Richard Falk, 1992, p. 6.
15. This is one of the central themes in David Osborne and Ted Gaebler's very important book *Reinventing Government,* 1992.
16. Thomas Berry, 1990, p. 134.
17. See the Foreword to Ilya Prigogine and Isabelle Stengers, 1984.
18. Anton C. Zijderveld, 1986, p. 65.

19. Christopher Lasch, 1979.
20. Ibid., p. 282.
21. Leo Esaki, 1984, pp. 2–5.
22. Nicholas Mosley, 1991, p. 188.
23. Werner Heisenberg, 1989, pp. 193–94.

_____ CHAPTER 2 LEARNING TO THINK THE IMPOSSIBLE _____

1. Richard Feynman, 1967.
2. Frank Shu, 1982, pp. 232–33.
3. H. Fröhlich, 1986, and Fritz-Albert Popp, 1986.
4. M. Rattemeyer and F.-A. Popp, 1981.
5. David Bohm, remark made during Oxford University seminar on "Mind and Matter," 1991.
6. Richard Pascale, 1991, p. 33.
7. Friedrich Nietzsche, 1964 (1882).
8. M. Merleau-Ponty, 1960, pp. 136–37.
9. John Burnet, 1963, p. 146. Quotes are fragments from Heraclitus and Plato.
10. Michael Redhead, 1987, p. 48.
11. Bernard d'Espagnat, p. 128.
12. David Bohm, 1951, p. 415.
13. Brigitte and Peter Berger, 1983, p. 154.
14. David Bohm, 1951, p. 414.
15. Peter Brook, 1988, p. 106.
16. S. Freedman and J. Clauser, 1972.
17. R. L. Pfleegor and L. Mandel, 1967.
18. M. Horne, A. Shimony, and A. Zeilinger, 1990, pp. 429–430.
19. David Bohm, 1980.

_____ CHAPTER 3 A NEW PHYSICS OF THE MIND _____

1. See, for instance, the discussion in Noam Chomsky, 1975.
2. Karl Popper and John Eccles, 1977.
3. John R. Searle, 1992.
4. Hubert L. Dreyfus, 1993.
5. Roger Penrose, 1989.
6. Jerry Fodor, 1991, p. 15.
7. Thomas Nagel, 1979, pp. 165–80.
8. Colin McGinn, 1991.
9. David Bohm, 1951, pp. 168–72.
10. Roger Penrose, 1987, p. 274.
11. Roger Penrose, 1989, pp. 444–45.
12. Descartes, "Sixth Meditation" (86), 1960, p. 81.
13. Jon Elster, 1989, p. 287.
14. See, for example, Evan Harris Walker, 1970; H. S. Green and T. Triffet,

1975; Lawrence Domash, 1976; Yuri Orlov, 1982; I. N. Marshall, 1989; Michael Lockwood, 1990; and David Hodgson, 1991. All present quantum-mechanical models of consciousness.

15. Roger Penrose, 1987, p. 274.

16. Roger Penrose, 1989, pp. 437–38.

17. David Bohm, 1980, pp. 143–47.

18. Ken Wilber, 1982, p. 2.

19. For a further discussion of Bose-Einstein condensation, see Tony Hey and Patrick Walters, 1987, pp. 115–18.

20. Dale Armin Miller, 1992, p. 364.

21. E. E. Witmer, 1966.

22. I. N. Marshall, 1989.

23. The most cogent argument for some kind of superconductivity in the brain was suggested by Evan Harris Walker, 1970.

24. H. Fröhlich, 1968.

25. E. Del Giudice, et al., 1989.

26. M. Rattemeyer and F.-A. Popp, 1981.

27. H. Fröhlich and F. Kremer, 1983; H. Fröhlich, 1986.

28. Fritz-Albert Popp, 1988, pp. 576–85.

29. Humio Inaba (Tohoku University Research Institute of Electrical Communication), 1989, p. 41.

30. Michael Stryker, 23 March 1989, p. 297.

31. Andreas K. Engel, et al., 1990, pp. 588–606.

32. R. Llinas, 1990, pp. 933–38.

33. Christine Skarda and Walter Freeman, 1987.

34. Chong Zheng, et al., 1988, p. 726.

35. V. I. Gol'danskii, et al., 1984, p. 209.

36. R. Llinas, 1990.

37. J. K. Tsotsos, 1990.

38. Roger Penrose, 1989, pp. 436–37.

39. C.M.H. Nunn, C.J.S. Clarke, and B. H. Blott, 1992.

40. H. Benson, 1975.

41. David Tank and John Hopfield, 1987, p. 16.

42. R. W. Doty, 1989.

43. I. N. Marshall, 1991, pp. 16–19.

44. Rodney Loudon, 1983, p. 281. Loudon points out that though, theoretically, the coherence time of a laser beam is one hundred seconds, in practice it is more like one second. The discrepancy is due to slight vibrations in the system that weaken the Bose-Einstein condensate.

―――――― CHAPTER 4 PRIVATE SELVES AND PUBLIC PERSONS ――――――

1. James Hillman, 1977, p. 42.

2. Don Bannister, referred to in "The Significance of the Signifier," *The New Scientist,* No. 1799, 14 December 1991, p. 50.

3. Charles Rycroft, 1968, pp. 100–102.

4. Aristotle, *On Democritus,* frag. 208, quoted in Jonathan Barnes, 1987, p. 247.

5. Thomas Hobbes, *Leviathan.*

6. Jon Elster, 1989, p. 287.

7. Norman P. Barry, 1989, pp. 71–73.

8. Quoted in Philip Rieff, 1972, p. 35.

9. Philip Rieff, 1972, p. 19.

10. Ibid., p. 22.

11. Parmenides (frags. 8, 25, and 45), cited in John Burnet, 1963 (1930), p. 145.

12. Jean-Jacques Rousseau, 1947 (1762), Chapter VI.

13. G.W.F. Hegel, 1948, p. 155.

14. Quoted in Peter Laslett, 1967, p. 279.

15. Leo Esaki, 1984, pp. 2–5.

16. Quoted in Jonathan Sacks, 1991, p. 13.

17. Peter Brook, *A Theatrical Casebook,* 1988, p. 340.

18. Ibid., p. 341.

19. David Osborne and Ted Gaebler, 1992, Chapter 9.

20. Emile Durkheim, *Rules of the Sociological Method,* quoted in Raymond Aron, 1970, p. 79.

21. John Rawls, 1972.

22. Thomas Nagel, 1991, Chapter 1.

23. See Paul Teller, 1986.

24. Emile Durkheim, *Elementary Forms of the Religious Life,* quoted in Raymond Aron, 1970, pp. 61–62.

25. William James, 1880. Quoted in *The Independent,* 5 August 1991, p. 17.

## CHAPTER 5 FREEDOM AND AMBIGUITY

1. Adam Smith, 1976 (1776), Vol. I, Book IV, Chapter II, p. 477.

2. John Stuart Mill, paraphrased by Isaiah Berlin, in Isaiah Berlin, 1969, p. 127.

3. Ibid., p. 478.

4. Robert N. Bellah, et al., 1991, p. 9.

5. See discussion in Michael Sandel, 1984, p. 9.

6. Robert N. Bellah, et al., 1991, p. 19.

7. Christopher Lasch, 1979.

8. Isaiah Berlin, 1969, p. 124.

9. Jonathan Sacks, 1991, pp. 63–64.

10. G.W.F. Hegel, quoted in Charles Taylor, 1984, p. 181.

11. Isaiah Berlin, 1975, pp. 131 and 137.

12. Martin Buber, 1937, p. 15.

13. Dale Armin Miller, 1992, pp. 361–62.

14. Freeman Dyson, 1959, pp. 249–50.

15. Dale Armin Miller, 1992, p. 364.

16. Peter Brook, 1968, p. 114.

17. Peter Berger 1984 (1970), p. 155.

_____ CHAPTER 6 THE MANY FACES OF TRUTH _____

1. Ernest Nagel and James R. Newman, 1959.
2. Plato, *Meno*, 81d, in *Collected Dialogues*, 1961, p. 364.
3. Descartes, *Discourse on Method*, [15], 1956, p. 10.
4. Isaiah Berlin, "Two Concepts of Liberty," 1990 (1969), p. 167.
5. David Harvey, 1990, pp. 35–36.
6. The best discussion I have seen of the fatal relation between instrumental reason and the Holocaust is in Zygmunt Bauman, 1989.
7. Charles Baudelaire, quoted in David Harvey, 1990, p. 10.
8. Kenneth Baynes, et al., 1987, p. 22.
9. Francis Fukuyama, 1992.
10. Richard Rorty, 1987, pp. 26–63.
11. Richard Rorty, 1979, p. 317.
12. David Lodge, 1989.
13. Alan Montefiore, Balliol College. Communicated in private conversation.
14. For a fuller discussion of consciousness as a creative self-organizing system, see Chapter 13 in *The Quantum Self*, Zohar 1990.
15. Chaim Potok, 1990, p. 344.
16. Michael Redhead, 1987, Chapter 2.
17. Chaim Potok, 1990, p. 70.
18. Peter Brook, 1988, p. 129.
19. Ilya Prigogine and Isabelle Stengers, 1984, Chapter VI.
20. Isaiah Berlin, 1969, p. 172.
21. Ibid.
22. T. S. Eliot, "Burnt Norton," *Four Quartets*.

_____ CHAPTER 7 A UNIVERSAL EVOLUTIONARY PRINCIPLE _____

1. Charles Darwin, 1860, p. 486.
2. Nicholas Mosley, 1991, p. 139.
3. Lee Smolin, 1992, pp. 173–91.
4. John Barrow and Frank Tipler, 1988, p. 16.
5. Ibid., p. 21.
6. This example is proposed in William Lane Craig, 1988, p. 392.
7. John Barrow and Frank Tipler, 1988, pp. 22–23.
8. Henri Bergson, 1964.
9. Alfred North Whitehead, 1978.
10. Teilhard de Chardin, 1959.
11. J. S. Bell, "Six Possible Worlds of Quantum Mechanics," in Bell, 1988, pp. 181–95.
12. See discussion in J. S. Bell, "Are There Quantum Jumps?" in Bell, 1988, pp. 202–204.
13. Roger Penrose, 1989, Chapter 8.
14. J. Cairns, 1988.

15. B. G. Hall, 1991.

16. T. S. Eliot, "East Coker," *Four Quartets,* 1952 (1943).


_____ CHAPTER 8 A COMMUNITY OF COMMUNITIES _____

1. Athenian Society, 1992.

2. William Rees-Mogg, 4 May 1992.

3. Arthur M. Schlesinger, Jr., 1992.

4. Charlisle Lyles, quoted in Studs Terkel, 1992.

5. Paul Ricoeur, 1961, p. 278.

6. Jonathan Sacks, 1991, p. 64.

7. Zygmunt Bauman, 1991, pp. 6 and 14.

8. Paraphrased by Dr. Malcolm Pines, a London-based Freudian analyst.

9. Frans de Waal, 1991, pp. 260–61.

10. Christine A. Skarda and Walter J. Freeman, June 1987, pp. 161–195.

11. Paul Ricoeur, 1961, p. 283.

12. Jonathan Sacks, 1991, p. 66.


_____ CHAPTER 9 THE FIFTH SON _____

1. John Kenneth Galbraith, "Why America Doesn't Care," *The Independent* (London), 8 May 1992. This article is an extract from his new book. See Galbraith, 1992.

2. Thich Nhat Hanh, 1987, pp. 62–64.

3. Robert N. Bellah, et al., 1991, p. 257.

4. Mihaly Csikzentmihalyi and Eugene Rochberg-Holton, 1981, Chapter 1. Cited in Robert N. Bellah, et al., 1991, p. 257.

5. John Keane, 1992, p. 10.

6. Ibid.

7. Robert N. Bellah, et al., 1991, p. 4.

8. Ibid., p. 12.

9. Ibid., p. 288.

10. C. G. Jung, "The Meaning of Psychology for Modern Man," p. 149.


_____ CHAPTER 10 THE SELF, THE VACUUM, AND THE CITIZEN _____

1. Personal communication from Maria Guarnaschelli.

2. Robert N. Bellah, et al., 1991, p. 292.

3. Plato, *Phaedo,* 66a–c.

4. Epistle to the Romans 7:24–25.

5. John Calvin, 1961 (1559), p. 246.

6. John Milton, "On the Morning of Christ's Nativity."

7. Genesis 1:28.

8. Quoted in Ian Bradley, 1990, p. 12. Bradley's book offers an excellent survey of Christian attitudes toward nature.

9. *First Meditation* 19 [14], in René Descartes, 1960.

10. Nicola Abbagnano, 1967, p. 70.

11. *Discourse on Method* [20], in René Descartes, 1961.

12. David Hume, (1748), Sec. XII, Part III.

13. See, for instance, David Hodgson's account of legal reasoning and the inability of rationalistic computer models of mind to account for it. David Hodgson, 1991.

14. Bede Griffiths, 1989, Chapter 13, "The New Age." There is also extensive discussion of the devastating effects of modern science on our wider spiritual concerns in Brian Appleyard, 1992.

15. Zygmunt Bauman, 1989, p. 18.

16. Gianni Vattimo, 1988, p. xxii.

17. Charlene Spretnak, 1991, p. 15.

18. Gianni Vattimo, p. xxv.

19. Iris Murdoch, 1985, p. 30.

20. Oxford philosopher John Gray makes this same point about the incompatibility of modern humanism and an ecological worldview. See John Gray, 1992.

21. This is pretty much a precis of the Green political program outlined in Edward Goldsmith's new book. See Edward Goldsmith, 1992.

22. Rodney Loudon, 1983, Chapter 6.

23. There is a discussion of the brain's many evolving layers in both *The Quantum Self*, Chapter 5, and Gerald Edelman, 1992, Chapter 14.

24. There are many references to and descriptions of the vacuum at an almost nontechnical level in G. D. Coughlan and J. E. Dodd, 1991.

25. Quoted in Heinrich Zimmer, 1951, pp. 520–22.

26. Heinz R. Pagels, 1982, p. 274.

27. Ibid., pp. 274–75.

28. Tony Hey and Patrick Walters, 1987, p. 151.

29. Ibid., p. 275.

30. Michael Serres, quoted in Ilya Prigogine and Isabelle Stengers, 1984, pp. 304–305.

31. Richard Falk, 1992, p. 36.

32. Arthur William Edgar O'Shaughnessy, "Ode."

33. Richard Falk, 1992, p. 16.

—— CHAPTER 11 THE POLITICS OF TRANSFORMATION I ——

1. Alasdair Macintyre, 1987 (1981), p. 2.

2. "Disenchantment in the West," *The Independent,* 19 October 1992, p. 18.

3. A telling example of this was offered by an American polling firm that contrasted voters' priorities with those of the 1988 presidential candidates. See William Greider, 1992, p. 22.

4. Philip Rieff, 1987, p. 93.

5. Richard Falk, 1992, p. 9.

6. Niccolò Machiavelli, 1981 (1514), XV.

7. Eric R. Kandel and Robert D. Hawkins, 1992.

8. The many characteristics of small groups I have mentioned here are agreed on in most books on group psychology. See for instance Rupert Brown, 1988, pp. 1–18.

9. David W. Tank and John J. Hopfield, 1987.

10. Eric R. Kandel and Robert D. Hawkins, 1992.

11. D. A. Norman, 1987.

12. Jerry Gray and Frederick Starke, 1984, p. 461.

13. Ibid., p. 461.

14. Robin Dunbar, 1992, p. 28.

15. There is a lengthy and fascinating discussion of the contrast between Western notions of private meaning and Japanese notions of public meaning and the implications for organization and management theory in Richard Tanner Pascale and Anthony Athos, 1982.

16. Jeremy Waldron, 1990, p. 715.

17. R. Wuthnow, 1986.

18. Jeremy Waldron, 1990, p. 715.

19. Gunnar Hjelholt, 1975.

20. Alasdair Macintyre, 1987 (1981), p. 34.

21. Robert Nozick, 1991 (1974), p. 26. A description of the "night watchman" theory of the minimal state.

22. Isaiah Berlin, 1991, p. 260.

23. Ibid., pp. 260–61.

24. Ibid., p. 247.

25. Ibid., p. 255.

26. John Keane, 1992, p. 10.

27. Ibid.

28. Ibid., p. 11.

29. Jonathan Sacks, 1992, pp. 4–5.

30. *The Independent,* 19 November 1992, p. 18.

31. Jonathan Sacks, 1992. The whole book is about this problem.

32. Robert N. Bellah, et al., 1991, p. 61.

—— CHAPTER 12 THE POLITICS OF TRANSFORMATION II ——

1. Peter Berger, 1967.

2. George Emery Mendenhall, 1981, p. 227.

3. Peter Berger, et al., 1974, p. 77.

4. Hans Küng, 1991.

5. James Hillman, 1981.

6. Philip Rieff, 1981, p. 98.

7. Eric R. Kandel and Robert D. Hawkins, 1987.

8. J. A. Gray, 1987.

9. Henri Matisse, in *Le Courier de l'Unesco,* October 1953.

10. David Bohm and F. David Peat, 1987, p. 240.

11. Ibid., p. 243.

12. Ibid.

13. Ibid., p. 244.

14. Ibid., p. 242.

15. Stuart Hampshire, 1991, p. 21.

16. Jerry Gray and Frederick Starke, 1984, pp. 533–43.

17. Ibid., p. 536.

18. Gerald Edelman, 1992, p. 29. There is extensive discussion of the evolving brain in this book.

19. Stuart Hampshire, 1991, p. 26.

20. David Bohm, 1990, p. 2.

21. Emilios Bouratinos, 1992.

22. Patrick de Mare, 1991.

23. Philip Rieff, 1987, p. 13.

24. *The Independent*, 28 November 1992.

25. David Osborne and Ted Gaebler, 1992. The whole book contains scores of examples.

26. Gerald Edelman, 1992, p. 29.

27. Richard Falk, 1992, p. 16.

28. Ibid., p. 72.

29. C. G. Jung, 1964, Vol. 10, para. 315.

## —————— CHAPTER 13 REINVENTING THE FAMILY ——————

1. Judith S. Wallerstein and Sandra Blakeslee, 1989, pp. 148–49, cited in *The Good Society*. This is only one of many such studies to follow up the fate of children whose families have broken up.

2. Robert N. Bellah, et al., 1991, pp. 256–57. Again, there are many similar studies that reveal these same facts.

3. Andrew Sims, 1992.

4. Judith S. Wallerstein and Sandra Blakeslee, 1989, p. xxi, cited in *The Good Society*.

5. Gershom Scholem, 1963, p. 35.

6. Peter Berger 1984 (1970), p. 149.

7. Peter Berger makes much of the distinction between individualist dignity and communal honor. See Peter Berger 1984 (1970), p. 154.

8. Robert N. Bellah, et al., 1991, p. 46.

9. Peter Berger 1984 (1970), p. 156.

10. Jerome Frank, 1975, p. 8.

11. A. Guggenbuhl-Craig, 1977, p. 34.

12. Robert N. Bellah, et al., 1991, p. 46.

13. Quoted earlier in Chapter 4. From Aristotle's *On Democritus*, frag. 208. Jonathan Barnes, 1987, p. 247.

14. This is actually an expression originating with the coauthor of this book, Dr. Ian Marshall.

15. Plato, *The Symposium*, 191e, in Plato, 1961.

16. Martin Buber, 1937.

17. E. Bruce Taub-Bynum, 1984.

18. Charles Rycroft, 1971, p. 66.

19. Ibid., p. 67.

20. The Dalai Lama in interview with William Rees-Mogg, the London *Times,* May 6, 1993, p. 18.

21. Robert N. Bellah, et al., 1991, p. 257.

## ——— CHAPTER 14 TOWARD QUANTUM CONSENSUS ———

1. Chris Johnston, 1993, "Society" section, *Guardian,* 10 February 1993, p. 13.

2. Rawls, Nozick, and Hampshire all have been quoted in earlier chapters of this book.

3. Thomas Kuhn, 1962.

4. Susanne K. Langer, 1952, Chapter 1.

5. Michael Walzer, 1992, pp. 66 and 67.

6. Ibid., p. 67.

7. Susanne K. Langer, 1952, p. 293.

8. Thomas Nagel, 1991.

9. Tracy B. Strong, 1992, p. 8.

# · Bibliography ·

Abbagnano, Nicola. "Humanism." In Paul Edwards, ed., *The Encyclopedia of Philosophy*. New York and London: Macmillan, 1967.

Appleyard, Brian. *Understanding the Present*. London: Pan, 1992.

Aron, Raymond. *Main Currents in Sociological Thought*, Vol. 2. London: Pelican, 1970.

Athenian Society for Science and Human Development. *II International Symposium on Science and Consciousness*. Athens: 4 Peta Street, Plaka, 10558 Athens, 1992.

Barnes, Jonathan. *Early Greek Philosophy*. London: Penguin, 1987.

Barrow, John D., and Tipler, Frank J. *The Anthropic Cosmological Principle*. Oxford and New York: Oxford University Press, 1988.

Barry, Norman P. *An Introduction to Modern Political Theory*. London: Macmillan, 1989.

Bauman, Zygmunt. *Modernity and the Holocaust*. Oxford: Polity Press, 1989.

Bauman, Zygmunt. "Postmodernity: Chance or Menace?" Lancaster University: Centre for the Study of Cultural Values, 1991.

Baynes, Kenneth, Bohman, James, and McCarthy, Thomas. *After Philosophy: End or Transformation*. Cambridge, Mass., and London: MIT Press, 1988.

Bell, J. S. *Speakable and Unspeakable in Quantum Mechanics*. Cambridge, New York, etc.: Cambridge University Press, 1988.

Bellah, Robert N., et al. *The Good Society*. New York: Alfred A. Knopf, 1991.

Benson, H. *The Relaxation Response*. New York: William Morrow, 1975.

Berger, Brigitte and Peter. *The War over the Family*. New York and London: Penguin, 1983.

Berger, Peter. *The Sacred Canopy*. Garden City, N.Y.: Doubleday, 1967.

Berger, Peter, Berger, Brigitte, and Kellner, Hansfried. *The Homeless Mind*. New York and London: Penguin, 1974.

Berger. Peter. "On the Obsolescence of the Concept of Honour." In Michael J. Sandel, ed., *Liberalism and Its Critics*. Oxford: Basil Blackwell, 1984 (1970).

343

Bergson, Henri. *Creative Evolution*. New York and London: Macmillan, 1964.

Berlin, Isaiah. *Four Essays on Liberty*. Oxford: Oxford University Press, 1990 (1969).

Berlin, Isaiah. *The Crooked Timber of Humanity*. London: Fontana, 1991.

Berry, Thomas. *The Dream of the Earth*. San Francisco: Sierra Club Books, 1990.

Blakemore, Colin. *The Mind Machine*. London: BBC Books, 1988.

Bloom, Allan. *The Closing of the American Mind*. New York and London: Simon and Schuster, 1987.

Bohm, David. *Quantum Theory*. London: Constable, 1951.

Bohm, David. *Wholeness and the Implicate Order*. London, Boston, and Henley, UK: Routledge & Kegan Paul, 1980.

Bohm, David. "A New Theory of the Relationship of Mind and Matter." *Philosophical Psychology*, Vol. 3, No. 2 (1990).

Bohm, David. *On Dialogue*. Ojai, Calif.: David Bohm Seminars, 1990.

Bohm, David, and Peat, F. David. *Science, Order and Creativity*. New York, London, etc.: Bantam, 1987.

Bourotinos, Emilios. "Interpersonal Democracy: The Quantum Approach." Proceedings of *Political Philosophy Today: Considerations After the End of Communism*. Crete: University of Crete, July 1992.

Bradley, Ian. *God Is Green*. London: Darton, Longman and Todd, 1990.

Brook, Peter. *The Empty Space*. London and New York: Penguin, 1968.

Brook, Peter. *The Shifting Point*. London: Methuen Drama, 1988.

Brook, Peter. *A Theatrical Casebook*. London: Methuen, 1988.

Brown, Rupert. *Group Processes*. Oxford: Basil Blackwell, 1988.

Buber, Martin. *I and Thou*. Edinburgh: T. & T. Clark, 1937.

Burnet, John. *Early Greek Philosophy*. Cleveland and New York: Meridian Books, 1963 (1930).

Cairns, John, et al. "The Origin of Mutants." *Nature*, Vol. 335, pp. 142–45 (8 September 1988).

Calvin, John. *Institutes of the Christian Religion*. Ed. J. T. McNeill. London: SCM Press, 1961 (1559).

Capra, Fritjof. *The Tao of Physics*. London: Fontana/Collins, 1976.

Carroll, Lewis. *Through the Looking Glass*. London: Macmillan, 1980.

Chomsky, Noam. "Recent Contributions to the Theory of Innate Ideas." In S. P. Stitch, ed., *Innate Ideas*. Berkeley: University of California Press, 1975.

Coughlan, G. D., and Dodd, J. E. *The Ideas of Particle Physics*. 2nd edition. Cambridge: Cambridge University Press, 1991.

Craig, William Lane. "Barrow and Tipler on the Anthropic Principle vs. Divine Design." *British Journal for the Philosophy of Science*, Vol. 38 (1988), pp. 389–95.

Csikzentmihalyi, Mihaly, and Rochberg-Holton, Eugene. *The Meaning of Things: Domestic Symbols and the Self*. Cambridge: Cambridge University Press, 1981.

Darwin, Charles. *On the Origin of Species*. 2nd edition. London: John Murray, 1860.

de Bono, Edward. *Handbook for the Positive Revolution*. New York and London: Viking, 1991.

de Mare, Patrick. *Koinonia: From Hate Through Dialogue to Culture in a Large Group*. London and New York: Karnac Books, 1991.

de Santillana, Giorgio. "Positivism and the Technocratic Ideal in the Nineteenth Century." In M. F. Ashley Montagu, ed., *Studies and Essays in the History of Science and Learning Offered in Homage to George Sarton*. New York: 1946, pp. 249–59.

de Waal, Frans. *Peacemaking Among Primates*. London and New York: Penguin, 1991.

Del Giudice, E., Doglia, S., Milani, M., Smith, C. W., and Vitiello, G. "Magnetic Flux Quantization and Josephson Behaviour in Living Systems." *Physica Scripta*, Vol. 40 (1989), pp. 786–91.

Del Giudice, E., Preparata, G., and Vitiello, G. "Water as a Free Electric Dipole Laser." *Physical Review Letters*, Vol. 61 (1988), pp. 1085–88.

Descartes, René. *Meditations*. Indianapolis and New York: Bobbs-Merrill Co., 1960.

Descartes, René. *Essential Works of Descartes*. New York: Bantam, 1961.

D'Espagnat, Bernard. *Reality and the Physicist*. Cambridge: Cambridge University Press, 1989.

Domash, Lawrence H. "The Transcendental Meditation Technique and Quantum Physics: Is Pure Consciousness a Macroscopic Quantum State in the Brain?" In David Orme-Johnson and John T. Farrow, eds., *Scientific Research on the Transcendental Meditation Program*, Vol. I. Geneva: Maharishi European Research University Press, 1976.

Doty, R. W. "Schizophrenia: A Disease of Interhemispheric Brain Processes at Forebrain and Brainstem Levels?" *Behavioural Brain Research*, Vol. 34 (1989), pp. 1–33.

Dray, W. H. "Holism and Individualism in History and Social Science." In Paul Edwards, ed., *The Encyclopedia of Philosophy*. London: Collier-Macmillan, 1967.

Dreyfus, Hubert L. *What Computers Still Can't Do*. New York: Harper and Row, 1993.

Dunbar, Robin. "Why Gossip Is Good for You." *The New Scientist*, 21 November 1992, pp. 28–31.

Dyson, Freeman. *Disturbing the Universe*. New York: Basic Books, pp. 249–50.

Edelman, Gerald. *Bright Air, Brilliant Fire*. New York and London: Penguin (Allen Lane), 1992.

Eliot, T. S. *Four Quartets*. London: Faber and Faber, 1952 (1943).

Elster, Jon. *The Cement of Society*. Cambridge: Cambridge University Press, 1989.

Elster, Jon. *Ulysses and the Sirens: Studies in Rationality and Irrationality*. Cambridge and New York: Cambridge University Press, 1984.

Engel, Andreas K., Konig, Peter, Gray, Charles M., and Singer, Wolf. "Stimulus-Dependent Neuronal Oscillations in Cat Visual Cortex: Inter-Columnar Interaction as Determined by Cross-Correlation Analysis." *European Journal of Neuroscience*, Vol. 2 (1990), pp. 588–606.

Esaki, Leo. "Why the West Misreads Japan." *The Bridge*, Journal of the National Academy of Engineering, Vol. 14, No. 1 (1984), pp. 2–5.

Falk, Richard. *Explorations at the Edge of Time*. Philadelphia: Temple University Press, 1992.

Feynman, Richard. *The Character of Physical Law*. Cambridge, Mass.: MIT Press, 1967.

Fodor, Jerry. " 'The Problem of Consciousness.' " *Times Literary Supplement,* 7 June 1991, p. 15.

Foley, Vernard. *The Social Physics of Adam Smith*. West Lafayette, Ind.: Purdue University Press, 1976.

Frank, Jerome D. "An Overview of Psychotherapy." In Gene Usdin, ed., *Overview of the Psychotherapies*. New York: Bruner/Mazel, 1975.

Freedman, S., and Clauser, J. "Experimental Test of Local Hidden Variables Theories." *Physical Review Letters,* Vol. 28 (1972).

Fröhlich, H. "Long-Range Coherence and Energy Storage in Biological Systems." *International Journal of Quantum Chemistry,* Vol. II (1968).

Fröhlich, H. "Coherent Excitations in Active Biological Systems." In F. Gutman and H. Keyzer, eds., *Modern Biochemistry*. New York and London: Plenum, 1986.

Fröhlich, H., and Kremer, F., eds. *Coherent Excitations in Biological Systems*. Berlin, Heidelberg, New York, and Tokyo: Springer-Verlag, 1983.

Fukuyama, Francis. *The End of History and the Last Man*. New York and London: Penguin, 1992.

Galbraith, John Kenneth. *The Culture of Contentment*. London: Sinclair-Stevenson, 1992.

Giddens, Anthony. *The Consequences of Modernity*. Stanford, Calif., and Oxford: Polity Press, 1990.

Gol'danskii, V. I., et al. "Tunneling Between Quasi-Degenerate Conformational States and Low-temperature Heat Capacity of Biopolymers. The Glasslike Protein Model." *Doklady Biophysics,* Vol. 272, No. 4 (October 1983).

Goldsmith, Edward. *The Way*. London, Sydney, Auckland, and Johannesburg: Rider, 1992.

Goudge, T. A. "Emergent Evolutionism." In Paul Edwards, ed., *The Encyclopedia of Philosophy*. New York and London: Macmillan, 1967, pp. 474–76.

Gray, J. A. *The Psychology of Fear and Stress*. Cambridge: Cambridge University Press, 1987.

Gray, Jerry L., and Starke, Frederick A. *Organizational Behavior*. Columbus, Toronto, etc.: Charles E. Merrill, 1984.

Gray, John. "Saving the Biosphere." *Times Literary Supplement,* 11 September 1992, p. 5.

Green, H. S., and Triffet, T. "Quantum Mechanics and the Brain." *International Journal of Quantum Chemistry:* Quantum Biology Symposium No. 2 (1975), 289–96.

Greider, William. *Who Will Tell the People?* New York, London, etc.: Simon and Schuster, 1992.

Griffiths, Bede. *A New Vision of Reality*. London: Collins, 1989.

Guggenbuhl-Craig, A. *Marriage—Dead or Alive*. Zurich: Spring, 1977.

Hall, B. G. "Adaptive Evolution That Requires Multiple Spontaneous Mutations: Mutations Involving Base Substitutions." *Proc. Nat. Acad. Sci. U.S.A.*, Vol. 88 (1991), pp. 5882–86.

Hampshire, Stuart. "Justice Is Strife." Proceedings of the American Philosophical Association, Vol. 65, No. 3 (November 1991).

Hanh, Thich Nhat. "Please Call Me by My True Names." *Being Peace*. Berkeley, Calif.: Parallax Press, 1987.

Harvey, David. *The Condition of Postmodernity*. Oxford: Basil Blackwell, 1990.

Hegel, G.W.F. *Early Theological Writings*. Trans. T. M. Knox. Chicago: University of Chicago Press, 1948.

Heisenberg, Werner. *Physics and Philosophy*. New York and London: Penguin, 1989.

Hendry, G. S. *Theology of Nature*. Philadelphia: Westminister Press, 1980.

Hey, Tony, and Walters, Patrick. *The Quantum Universe*. New York, London, etc.: Cambridge University Press, 1987.

Hillman, James. *Re-Visioning Psychology*. New York, London, etc.: Harper Colophon, 1977.

Hillman, James. *Archetypal Psychology*. Dallas: Spring Publications, Inc., 1981.

Hjelholt, Gunnar. "Europe Is Different: Boundary and Identity as Key Concepts." In Hofstede and Kassem, eds., *European Contributions to Organization Theory*. Amsterdam: Assen, 1975, pp. 232–43.

Hobbes, Thomas. *Leviathan*. Glasgow: Collins Fount, 1983 (1651).

Hodgson, David. *The Mind Matters*. Oxford: Oxford University Press, 1991.

Horne, Michael, Shimony, Abner, and Zeilinger, Anton. "Two Particle Interferometry." *Nature*, Vol. 347 (4 October 1990), pp. 429–30.

Hume, David. *An Inquiry Concerning Human Understanding*. New York: Library of Liberal Arts, 1955 (1748).

Inaba, Humio. Research reported in *The New Scientist*, 27 May 1989.

*The Independent*. "The Foundations Wobble." London: Independent Newspapers, 28 November 1992, p. 16.

Jammer, M. "Mass." In Paul Edwards, ed., *The Encyclopedia of Philosophy*. London: Collier-Macmillan, 1967.

Jencks, Charles. "The Post-Modern Agenda." In Charles Jencks, ed., *The Post-Modern Reader*. London: Academy; New York: St. Martin's Press, 1992.

Johnston, Chris. "Behind the Mask and the Motive." *The Guardian* (London), 10 February 1993, Society section, p. 13.

Jung, C. G. "The Meaning of Psychology for Modern Man." In *Civilization in Transition*, Vol. 10, *The Collected Works*. London: Routledge & Kegan Paul, 1964 (1934).

Jung, C. G. "The Spiritual Problem of Modern Man" (1931). In *Civilization in Transition*, Vol. 10, *The Collected Works*. London: Routledge & Kegan Paul, 1964.

Keane, John. "Democracy's Poisonous Fruit." *Times Literary Supplement*, 12 August 1992, pp. 10–12.

Koyre, Alexander. *Newtonian Studies*. Chicago: University of Chicago Press, 1968.

Kuhn, Thomas S. *The Structure of Scientific Revolutions.* Chicago and London: University of Chicago Press, 1962.

Küng, Hans. *Global Responsibility.* London: SCM, 1991.

Langer, Susanne K. *Philosophy in a New Key.* Cambridge, Mass.: Harvard University Press, 1957.

Lasch, Christopher. *The Culture of Narcissism.* New York: Warner Books, 1979.

Laslett, Peter. "The General Will." In Paul Edwards, ed., *The Encyclopedia of Philosophy,* Vol. 3. London: Collier-Macmillan, 1967.

Llinas, R. "Intrinsic Electrical Properties of Nerve Cells and Their Role in Network Oscillation." *Cold Spring Harbor Symposia on Quantitative Biology,* Vol. LV, pp. 933–38. Cold Spring Harbor, N.Y.: Cold Spring Harbor Press, 1990.

Lockwood, Michael. *Mind, Brain and the Quantum.* Oxford: Basil Blackwell, 1990.

Lodge, David. *Nice Work.* London: Penguin, 1989.

Loudon, Rodney. *The Quantum Theory of Light.* Oxford: Clarendon Press, 1983.

Machiavelli, Niccolò. *The Prince.* New York and London: Penguin, 1981 (1514).

Macintyre, Alasdair. *After Virtue.* London: Duckworth (1981) 1987.

Mansbridge, Jane J. *Beyond Self-Interest.* Chicago and London: University of Chicago Press, 1990.

Marshall, I. N. (Ninian). "ESP and Memory: A Physical Theory." *British Journal for the Philosophy of Science,* Vol. X, No. 40 (1960).

Marshall, I. N. "Consciousness and Bose-Einstein Condensates." *New Ideas in Psychology,* Vol. 7, No. 1 (1989).

Marshall, I. N. "Some Phenomenological Implications of a Quantum Model of Consciousness." *Where Does 'I' Come From?* Buffalo: State University of New York, Department of Cognitive Science, Conference Proceedings, 1991.

Marshall, I. N. "Towards a New Science of Consciousness." *II International Symposium on Science and Consciousness.* Athens: The Athenian Society for Science and Human Development, 4 Peta Street, Plaka, 10558 Athens, Greece, 1992, pp. 104–110.

Marshall, I. N. "A Unified Dynamics of Consciousness and Quantum Collapse." Submitted for publication, 1993.

McGuinn, Colin. *The Problem of Consciousness.* Oxford: Basil Blackwell, 1991.

Mendenhall, George Emery. "Covenant." *Encyclopedia Britannica.* Chicago, etc.: Encyclopedia Britannica Inc., 1981, Vol. 5, pp. 226–30.

Merleau-Ponty, M. "Le Philosophe et la Sociologie." *Éloge de la Philosophie.* Paris: Collection Idées, Gallimard, 1960.

Miller, Dale Armin. "Agency as a Quantum-theoretic Parameter—Synthethic and Descriptive Utility for Theoretical Biology." *Nanobiology,* Vol. 1 (1992), pp. 361–71.

Milton, John. "On the Morning of Christ's Nativity." In *Fifteen Poets.* Oxford: Clarendon Press, 1941.

Monod, Jacques. *Chance and Necessity.* New York: Vintage Books, 1972.

Mosley, Nicholas. *Hopeful Monsters.* London: Minerva, 1991.

Murdoch, Iris. *The Good Apprentice*. London: Penguin, 1985.

Nagel, Thomas. "What It Is Like to Be a Bat." In *Mortal Questions*. Cambridge: Cambridge University Press, 1979.

Nagel, Thomas. *Equality and Partiality*. Oxford: Oxford University Press, 1991.

Nietzsche, Friedrich. *Joyful Wisdom*. New York: Frederick Ungar, 1964 (1882).

Norman, D. A. "Reflections on Cognition and Parallel Distributed Processing." In James L. McClelland, et al., *Parallel Distributed Processing*, Vol. 2., Cambridge, Mass., and London: MIT Press, 1987.

Nozick, Robert. *Anarchy, State, and Utopia*. Oxford: Basil Blackwell, 1991 (1974).

Nunn, C.M.H., Clarke, C.J.S., and Blott, B. H. "Consciousness and Quantum Mechanics: A Pilot Experiment." University of Southampton. Submitted for publication, 1993.

Obukhov, S. P. "Self-Organized Criticality: Goldstone Modes and Their Interactions." *Physical Review Letters*, Vol. 65, No. 12 (17 September 1990), pp. 1395–98.

Orlov, Yuri F. "The Wave Logic of Consciousness: A Hypothesis." *International Journal of Theoretical Physics*, Vol. 21, No. 1 (1982).

Osborne, David, and Gaebler, Ted. *Reinventing Government*. Reading, Mass., etc: Addison-Wesley, 1992.

Pagels, Heinz R. *The Cosmic Code: Quantum Physics as the Language of Nature*. London: Michael Joseph, 1982.

Pareto, Vilfredo. *The Mind and Society*. New York: Harcourt Brace, 1935.

Pascale, Richard Tanner. *Managing on the Edge*. New York and London: Penguin Books, 1991.

Pascale, Richard Tanner, and Athos, Anthony. *The Art of Japanese Management*. London and New York: Penguin Books, 1982.

Penrose, Roger. "Minds, Machines and Mathematics." In Colin Blakemore and Susan Greenfield, eds., *Mindwaves*. Oxford: Basil Blackwell, 1987.

Penrose, Roger. *The Emperor's New Mind*. Oxford, New York, Melbourne: Oxford University Press, 1989.

Perls, Fritz. *Gestalt Therapy Verbatim*. New York: Bantam, 1969.

Pfleegor, R. L., and Mandel, L. "Interference of Independent Photon Beams." *Physical Review*, Vol. 159, No. 5 (25 July 1967).

Plato, *The Collected Dialogues*, ed. Edith Hamilton and Huntingdon Cairns. Bollingen Series LXXI. New York: Pantheon Books, 1961.

Popp, Fritz-Albert. "On the Coherence of Ultraweak Photon-emission from Living Tissue." In C. W. Kilmister, ed., *Disequilibrium and Self-Organization*. Dordrecht and Boston: D. Reidel Publishing Co., 1986.

Popp, Fritz-Albert, et al. "Physical Aspects of Biophotons." In *Experientia*, Vol. 44. Basel: Birkhauser Verlag, 1988.

Popper, Karl R., and Eccles, John C. *The Self and Its Brain*. Berlin, London, and New York: Springer-Verlag, 1977.

Potok, Chaim. *The Gift of Asher Lev*. London: Heinemann, 1990.

Prigogine, Ilya, and Stengers, Isabelle. *Order Out of Chaos*. New York, London, etc.: Bantam Books, 1984.

Rattemeyer, M., et al. "Evidence of Photon Emission from DNA in Living Systems." *Naturwissenschaften*, Vol. 68, No. 5 (1981).

Rawls, John. *A Theory of Justice*. Oxford: Oxford University Press, 1972.

Redhead, Michael. *Incompleteness, Nonlocality, and Realism*. Oxford: Clarendon Press, 1987.

Rees-Mogg, William. "The Sheriff Fiddles While the Town Burns." *The Independent* (London), 4 May 1992, p. 17.

Ricoeur, Paul. "Universal Civilization and National Cultures." In *History and Truth*. Evanston, Ill.: Northwestern University Press, 1965 (1961).

Rieff, Philip. *Fellow Teachers*. Chicago: University of Chicago Press, 1972.

Rieff, Philip. *The Triumph of the Therapeutic*. Chicago: University of Chicago Press, 1987.

Rogers, Carl. *Encounter Groups*. New York and London: Allen Lane, 1969.

Rorty, Richard. *Philosophy and the Mirror of Nature*. Princeton, N.J.: Princeton University Press, 1979.

Rorty, Richard. "Pragmatism and Philosophy." In Kenneth Baynes, et al., eds., *After Philosophy: End or Transformation*. Cambridge, Mass., and London: MIT Press, 1988.

Rousseau, Jean-Jacques. *The Social Contract*. New York: Hafner Publishing Co., 1947 (1762).

Russell, Bertrand. "A Free Man's Worship." In *Mysticism and Logic*. New York: Doubleday Anchor, 1957.

Rycroft, Charles. *Pyschoanalysis Observed*. London: Constable, 1966.

Rycroft, Charles. *Reich*. London: Fontana, 1971.

Sacks, Jonathan. *The Persistence of Faith*. (The Reith Lectures 1990). London: Weidenfeld and Nicolson, 1991.

Sacks, Jonathan. *Crisis and Covenant*. Manchester: Manchester University Press, 1992.

Sandel, Michael, ed. *Liberalism and Its Critics*. Oxford: Basil Blackwell, 1984.

Schlesinger, Arthur M., Jr. *The Disuniting of America*. New York and London: W. W. Norton & Co., 1992.

Scholem, Gershom G., ed. *The Zohar*. New York: Schocken Books, 1963.

Searle, John R. *The Rediscovery of the Mind*. Cambridge, Mass. and London: MIT Press, 1992.

Sennett, Richard. *The Conscience of the Eye: The Design and Social Life of Cities*. London: Faber and Faber, 1991.

Sennett, Richard. "The body and the city." *Times Literary Supplement*, 18 September 1992, p. 3.

Serres, Michel. *Hermes: Literature, Science, Philosophy*. Baltimore and London: Johns Hopkins University Press, 1983.

Shanon, Benny, and Atlan, Henri. "Von Foerster's Theorem on Connectedness and

Organization: Semantic Applications." *New Ideas in Psychology,* Vol. 8, No. 1 (1990), pp. 81–90.

Sheldrake, Rupert. *The Presence of the Past.* London: William Collins, 1988.

Shu, Frank H. *The Physical Universe.* Mill Valley, Calif.: University Science Books, 1982.

Sims, Andrew. "Marital Breakdown and Health." *British Medical Journal,* Vol. 304 (22 February 1992), pp. 457–58.

Skarda, Christine A., and Freeman, Walter J. "How Brains Make Chaos in Order to Make Sense of the World." *Behavior and Brain Science,* Vol. 10 (1987), pp. 161–95.

Smith, Adam. *The Wealth of Nations.* Edited and with an introduction by Edwin Cannan. Chicago: University of Chicago Press, 1976 (1776).

Smith, Adam. *The Wealth of Nations.* Books I–III. Introduction by Andrew Skinner. London: Penguin, 1986 (1776).

Smith, Cyril, and Best, Simon. *Electromagnetic Man.* London: J. M. Dent & Sons Ltd., 1989.

Smolin, Lee. "Did the Universe Evolve?" *Classical and Quantum Gravity,* Vol. 9 (1992), pp. 173–91.

Spretnak, Charlene. *States of Grace: The Recovery of Meaning in the Postmodern Age.* San Francisco: HarperCollins, 1991.

Steinsaltz, Adin. *The Thirteen Petalled Rose.* New York: Basic Books, 1980.

Strong, Tracy B. "The Self and the Political Order." In Tracy B. Strong, ed., *The Self and the Political Order.* Oxford, UK, and Cambridge, Mass.: Basil Blackwell, 1992.

Stryker, Michael P. "Is grandmother an oscillation?" *Nature,* Vol. 338 (23 March 1989), pp. 297–98.

Tank, David W., and Hopfield, John J. "Collective Computation in Neuronlike Circuits." *Scientific American,* Vol. 257, No. 6 (December 1987).

Taub-Bynum, E. Bruce. *The Family Unconscious.* Wheaton, Ill.: Theosophical Publishing House, 1984.

Teilhard de Chardin, Pierre. *The Phenomenon of Man.* London: Collins, 1959.

Teller, Paul. "Relational Holism and Quantum Mechanics." *British Journal for the Philosophy of Science,* Vol. 37 (1986).

Terkel, Studs. *Race.* London: Sinclair-Stevens, 1992.

Thomas, Lewis. "The Lives of a Cell." *The Wonderful Mistake: Notes of a Biology Watcher.* Oxford: Oxford University Press, 1988.

Tilby, Angela. *Let There Be Light.* London: Darton, Longman and Todd, 1989.

Toynbee, Arnold J. *A Study of History.* Abridged by D. C. Somervell. London, New York, Toronto: Oxford University Press, 1951.

Tsotsos, J. K. "Analyzing Vision at the Complexity Level." *Behavioural and Brain Sciences,* Vol. 13 (1990), pp. 423–69.

Vattimo, Gianni. *The End of Modernity.* Oxford: Polity Press, 1988.

Waldron, Jeremy. "Politics Without Purpose?" *Times Literary Supplement,* 6 July 1990, pp. 715–16.

Walker, Evan Harris. "The Nature of Consciousness." *Mathematical Biosciences,* Vol. 7 (1970).

Wallerstein, Judith S., and Blakeslee, Sandra. *Second Chances: Men, Women and Children a Decade After Divorce.* New York: Ticknor and Fields, 1989.

Walzer, Michael. "On the Role of Symbolism in Political Thought." In Tracy B. Strong, ed., *The Self and the Political Order.* Oxford, UK, and Cambridge, Mass.: Basil Blackwell, 1992.

Weber, Eugene. "The Pleasures of Diversity." *Times Literary Supplement,* 22 August 1986, p. 906.

Weber, Max. *From Max Weber: Essays in Sociology.* Ed. H. H. Gerth and C. Wright Mills. London: Routledge, 1991.

Whitehead, Alfred North. *Process and Reality.* New York and London: Macmillan "Freepress," 1978.

Wilber, Ken, ed. *The Holographic Paradigm and Other Paradoxes.* Boulder, Colo., and London: New Science Library: 1982.

Witmer, E. E. "Interpretation of Quantum Mechanics and the Future of Physics." *American Journal of Physics,* Vol. 35 (1966), pp. 40–52.

Woodward, C. Vann. "Freedom and the Universities." *The New York Review of Books,* Vol. XXXVIII, No. 13 (July 18, 1991), p. 33.

Wu, T. M., and Austin, S. "Bose-Einstein Condensation in Biological Systems." *Journal of Theoretical Biology,* Vol. 71 (1978), pp. 209–214.

Wuthnow, R. "Religion as a Sacred Canopy." In James Davison Hunter and Stephen C. Ainlay, eds., *Making Sense of Modern Times.* London and New York: Routledge & Kegan Paul, 1986.

Yeats, William Butler. "The Second Coming," 1921.

Zheng, Chong, et al. "Quantum Simulation of Ferrocytochrome C." *Nature,* Vol. 334 (25 August 1988), pp. 726–28.

Zijderveld, Anton C. "The Challenges of Modernity." In James Davison Hunter and Stephen C. Ainlay, eds., *Making Sense of Modern Times.* London and New York: Routledge & Kegan Paul, 1986.

Zimmer, Heinrich. *Philosophies of India.* New York: Bollingen Foundation, Pantheon Books, 1951.

Zohar, Danah. *The Quantum Self.* London: HarperCollins/Flamingo, 1991; New York: William Morrow/Quill, 1991.

Zukav, Gary. *The Dancing Wu Li Masters.* London: Rider/Hutchinson, 1979.

# • Index •

Printed in the United States
1234000003B/61